AN INTRODUCTION TO ELECTROCHEMICAL SCIENCE

THE WYKEHAM SCIENCE SERIES

The aim of the Wykeham Science Series is to introduce the present state of the many fields of study within science to students approaching or starting their careers in University, Polytechnic, or College of Technology. Each book seeks to reinforce the link between school and higher education, and the main author, a distinguished worker or teacher in the field, is assisted by an experienced sixth form schoolmaster.

AN INTRODUCTION TO ELECTROCHEMICAL SCIENCE

J. O'M. Bockris
Flinders University of South Australia
N. Bonciocat
Academy of Science, Bucharest
F. Gutmann
University of Sydney

WYKEHAM PUBLICATIONS (LONDON) LTD
(A MEMBER OF THE TAYLOR & FRANCIS GROUP)
LONDON AND WINCHESTER
1974

First published 1974 by Wykeham Publications (London) Ltd.

Cover illustration The hydrogen-oxygen fuel cell inside the base of the Gemini space capsule

ISBN 0 85109 410 4 (paper)
ISBN 0 85109 420 1 (cloth)

Printed in Great Britain by Taylor & Francis Ltd.
10–14 Macklin Street, London, WC2B 5NF

Distribution:

UNITED KINGDOM, EUROPE, MIDDLE EAST AND AFRICA
Chapman & Hall Ltd. (a member of Associated Book Publishers Ltd.), 11 New Fetter Lane, London, EC4P 4EE, and North Way, Andover, Hampshire.

WESTERN HEMISPHERE
Springer-Verlag New York Inc., 175 Fifth Avenue, New York, New York 10010.

AUSTRALIA AND NEW GUINEA
Hicks Smith & Sons Pty. Ltd., 301 Kent Street, Sydney, N.S.W. 2000.

NEW ZEALAND AND FIJI
Hicks Smith & Sons Ltd., 238 Wakefield Street, Wellington.

ALL OTHER TERRITORIES
Taylor & Francis Ltd., 10–14 Macklin Street, London, WC2B 5NF.

PREFACE

Electrochemical Science deals with effects that involve electron transfer between two phases, one of which is electronically conducting and the other ionically conducting.

The readers we have in mind are those in the higher forms of secondary schools, and in the lower levels of universities and polytechnics. Electrochemical Science deals with the relations between the flow of electricity and the chemical constitution of matter, as well as with the electrical energy which can be derived directly from chemical reactions. And because of the increasing pollution caused by the use of fossil fuels to produce energy and the use of chemical technology to produce materials it may well be that the continuation of an affluent civilization (if this is what we want) into the next century will be difficult unless we change to clean electrically based technologies.

Electrochemical Science shows the way forward in this direction, and already indicates some of the changes that should be made.

J. O'M. Bockris, Adelaide, Australia
N. Bonciocat, Bucharest, Rumania
F. Gutmann, Sydney, Australia

CONTENTS

PRINCIPAL SYMBOLS

ΔG	free energy change
ΔH	heat change
ϵ_o	permittivity of a vacuum ($8{\cdot}854 \times 10^{-12}$ F m^{-1})
ϵ_r	relative permittivity
$\epsilon = \epsilon_o \epsilon_r$	permittivity
z_i	valency of ion
P	force, usually electrostatic
f_i	activity coefficient of ion
$f_\pm$	mean activity coefficient of two or more ions making up electrolyte
M	unit of molar concentration, in mol dm^{-3} (=mol l^{-1})
n_h	hydration number
I	ionic strength; also used for electric current.
V_{rev}	thermodynamically reversible potential of an electrode
$V_{rev}{}^0$	standard electrode potential
i_{diff}	diffusion current density
D	diffusion coefficient
$t_+ t_-$	transport numbers of cations and anions respectively
$\tilde{\mu}_i$	chemical potential of species i
μ_i	electrochemical potential of species i
V	potential difference
ψ	potential in double layer
κ	reciprocal of double layer thickness
E_F	Fermi level of electrons in a metal
i	current density (unit, ampere per square metre of electrode surface)
β	symmetry factor
i_+	partial anodic current density
i_-	partial cathodic current density
i_0	exchange current density
η	overpotential
V_{corr}	corrosion potential (sometimes called mixed, or compromise, potential)
ϵ_{max}	intrinsic maximum efficiency of fuel cell
ϵ_{real}	actual efficiency of fuel cell
e	the magnitude of the charge on electron or proton $\simeq 1{\cdot}6 \times 10^{-19}$ C

e	an electron
h	the Planck constant $= 6{\cdot}626 \times 10^{-34}$ J s
k	the Boltzmann constant $= 1{\cdot}38 \times 10^{-23}$ J K^{-1}
R	the gas constant $= N_A k = 8{\cdot}314$ J K^{-1} mol^{-1}
N_A	the Avogadro constant $= 6{\cdot}022 \times 10^{23}$ mol^{-1}
F	the Faraday constant $= N_A e = 9{\cdot}649 \times 10^{4}$ C mol^{-1}
eV	electron volt $= 1{\cdot}6 \times 10^{-19}$ J

CHAPTER 1

electrochemical science

1.1. *How we get our energy*

PRACTICALLY all the energy used by man at present comes to him in a way which, although we are used to it, on close examination appears quite extraordinary.

We start with a fossil fuel, for example, decane, which is burnt with oxygen from the air in, say, the cylinder of an internal combustion engine. The products of the reaction, carbon dioxide and water, are energetically more stable than the original substances, decane and oxygen. The energy difference between the reactants and products is given off as heat; the heat of combustion of decane is 6730 kJ mol^{-1} at atmospheric pressure and 293 K. The heated gaseous products expand and force a piston to move. The piston drives a wheel or, if electrical energy is needed, it turns the armature of a generator.

Continental areas	Average personal rate of consumption of energy/kW
North America	8·35
Oceania	3·31
Western Europe	2·77
Caribbean America	0·85
Other America	0·54
Western Asia	0·48
Far East	0·30
Africa	0·27
Other countries	1·33
World average	1·46

Table 1.1. World power consumption by continental areas (1966).

This (our usual) method of producing energy is indirect and complicated. There are two other negative aspects. One product (carbon dioxide) and several incidental by-products of the combustion reaction (complex, unsaturated hydrocarbons†, sulphur dioxide‡, and nitrogen

† The reaction of decane with O_2 does not yield only CO_2 and H_2O. Parallel reactions also occur.

‡ Because all fossil fuels contain sulphur; coal contains 1–5 per cent.

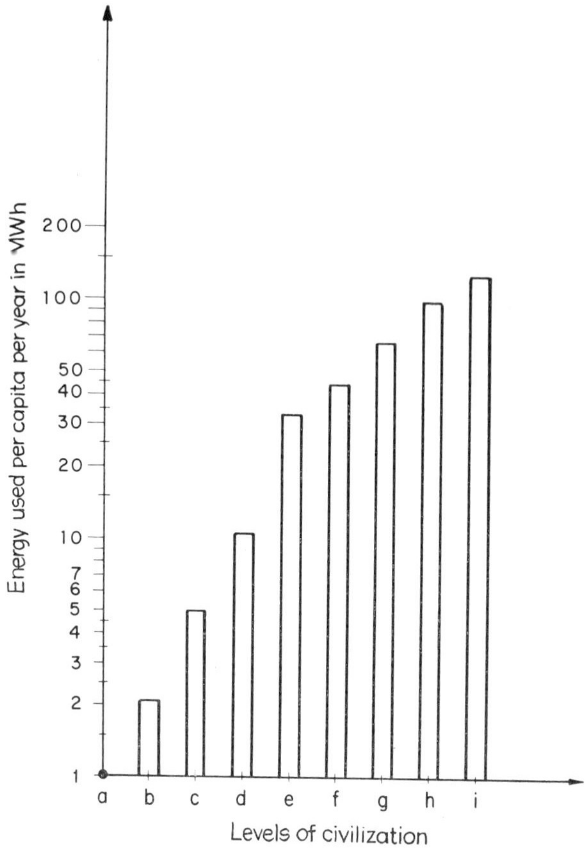

Fig. 1.1. Energy used per capita per year : *Data from* : C. Starr, *Scientific American*, **225** (3) (1971) 36, and E. Cook, *Scientific American*, **225** (3) (1971) 135. (*a*) Primitive man (1 000 000 years ago) ; (*b*) Hunting man (100 000 years ago) ; (*c*) Primitive agricultural man (5000 B.C.) ; (*d*) Advanced agricultural man (1400) ; (*e*) Industrial man (England, 1875) ; (*f*) U.S.A. (1900) ; (*g*) U.S.A. (1950) ; (*h*) U.S.A. (1970) ; (*i*) U.S.A. (2000, projected figure). (1 MWh $= 3{\cdot}6 \times 10^9$ J.)

oxides†) are air pollutants. With nitric oxide, and under the influence of light, the unsaturated hydrocarbon products form complicated addition compounds which condense on dust particles to form smog (fig. 1.2).

Secondly, internal combustion engines are very inefficient : two-thirds of the energy released is wasted as heat. External combustion (e.g. in the Stirling engine) is only slightly better.

† From the reaction of N_2 with O_2 in the hot cylinder.

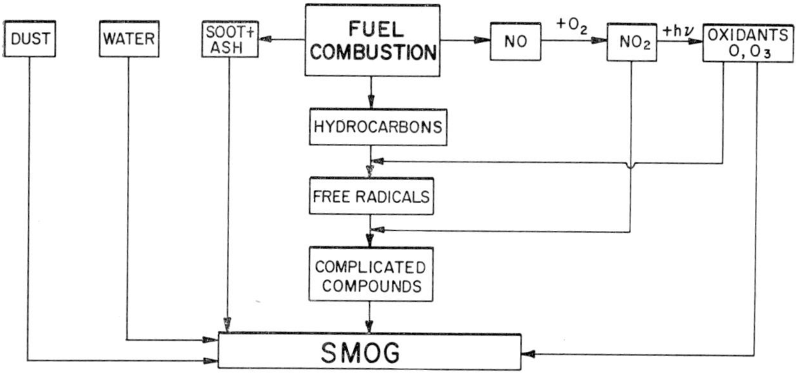

Fig. 1.2. The formation of smog (simplified).

1.2. *The end of the fossil fuel supply*

Such a way of obtaining energy might just be disregarded as a historical accident. However, an indisputable reason why we must explore other means is that the fossil fuels will not last for many more decades at acceptable cost.

Table 1.2 shows the known and hypothetical reserves. We have about one generation to go to the end of our known supply of fossil fuels. The hypothetical reserves (undersea oil; shale oil), which might (accepting the pollution they would cause) carry us perhaps a couple of generations more, are unreliable. No economically acceptable process for separating oil from shale has yet been developed.

Reserves	Life expectancy/year	
	Consumed at 1 Q per year (expected rate by 2000 A.D.)	Consumed at 2·5 Q per year (expected rate by 2050 A.D.)
Known	22	9
Hypothetical	462	177
Total	484	186

1 $Q = 1{\cdot}05 \times 10^{18}$ kJ, equivalent to $3{\cdot}8 \times 10^{12}$ kg of coal.

Table 1.2. Life expectancy of known and hypothetical reserves of fossil fuels (coal, oil, gas).

1.3. *Consequences of the presence of the products of combustion in the atmosphere*

Chimneys from power stations, and exhausts from cars, release the following pollutants: unsaturated hydrocarbons, CO, SO_2, NO, CO_2 and also a lead oxide.

Unsaturated hydrocarbons cause smog, bronchitis and cancer. Higher concentrations of carbon monoxide can cause fainting, and even death, because it combines with the haemoglobin of the blood. Nitrogen monoxide acts similarly. Sulphur dioxide gives an acrid smell and causes eye irritation.

The most insidious contaminant is carbon dioxide. The amount which cars and factories produce is greater than the amount which green plants can convert back to oxygen by photosynthesis. The concentration of carbon dioxide in the atmosphere has been increasing since about 1900 (fig. 1.3).

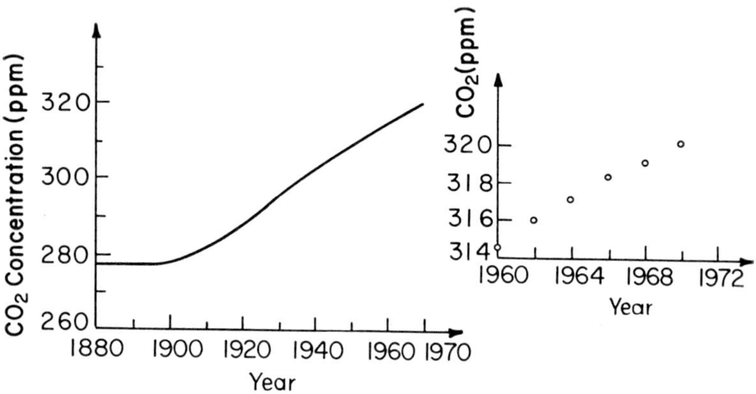

Fig. 1.3. The concentration (parts per million) of CO_2 in the atmosphere vs. time. After 1900, the balance between the rate of production and consumption of CO_2 has been disturbed by our fossil–fuel economy.

Carbon dioxide absorbs radiation, particularly the long infra-red. When the earth has been heated by the sun, it re-radiates as a black body at about 300 K, and it is the re-radiated energy which heats up the atmosphere. It can be predicted that some of the land-ice in the polar regions will in due course melt with a consequent rise in world sea level. The rise by A.D. 2000 might be about 30 cm, from then on the rise may be exponential.†

1.4. *Consequences of the rejection of used materials into seas, lakes and rivers*

Beginning about 1969, there has come to us the realization that we are killing our waters. Big bodies of fresh water, for example parts of the Great Lakes, the Hudson River, the Rhine, are now so polluted that their fish have largely died. Even the sea itself shows signs of pollution.

† This prediction would be upset for some years if sufficient solid matter in the atmosphere, e.g. soot or dust from combustion, reduces the intensity of the light reaching the earth's surface.

Why? The standard of living rises with the amount of energy available per capita. But each unit of energy used causes a material result, either a rejected product (e.g., detergent suds, rubbish, metal oxides) or a pollutant arising from the present way of producing the energy. Our high standard of life leads (with the methods we use at present to attain it) to a spoiling of our surroundings, which is tending to make affluent life repulsive.

1.5. *The need for recycling everything*

Consider water. We drink it, excrete it, our sun causes its evaporation, it rains, and so on.

The kind of recycling Nature provides with water should give us our clue. We do not need to exhaust materials, spoiling our environment and ending up with nothing to continue with. What we need to do is to treat each material in a way similar to that in which Nature treats water: we should use it, then supply energy to its used (degraded) form to upgrade it to its original state.

1.6. *The inexhaustible nature of clean energy*

Clean energy could be available to us in large quantities. The trouble is that, except for one source of energy—nuclear—we are not doing the research and development work at a sufficient intensity to put such energy at our disposal.

Three sources of inexhaustible energy are available:

(i) *Atomic energy*. When a nuclear reaction occurs, the amount of heat energy released per mole consumed is about 10^6 times the amount given out when 1 mole of a *chemical* undergoes transformation.

Huge amounts of energy are latent within the atom, and can be used to drive machines and produce electrical energy without the pollution entailed by burning coal or oil.

The deuterium–tritium reaction, based on the following reaction between deuterium and tritium nuclei

$$^{2}_{1}\mathrm{H} + ^{3}_{1}\mathrm{H} \rightarrow ^{4}_{2}\mathrm{He} + ^{1}_{0}\mathrm{n} + 17{\cdot}6\ \mathrm{MeV},\dagger$$

has been considered attractive because the mixture has the lowest known temperature of ignition to nuclear fusion (40×10^6 K). Such a reaction could be made to produce electricity by absorbing the heat energy released in a mantle of liquid metal (e.g. lithium). The hot metal would then be made to transfer its heat energy to water, generate steam and so to drive a turbine coupled to an electric generator.

Richard Post has proposed an alternative scheme for direct-energy conversion to electricity from the fusion process. In, for example,

$$^{2}_{1}\mathrm{H} + ^{3}_{2}\mathrm{He} \rightarrow ^{4}_{2}\mathrm{He} + ^{1}_{1}\mathrm{H} + 18{\cdot}3\ \mathrm{MeV}\dagger$$

† 1 MeV $= 10^6 \times 1{\cdot}6 \times 10^{-19}$ J $= 1{\cdot}6 \times 10^{-13}$ J. Multiplying this by $N_A = 6 \times 10^{23}$ gives for the molar heat of the reaction about 10^8 kJ.

the positively charged reaction products—from the plasma produced in the fusion reactor—are collected at a series of electrodes, and the electrons from the plasma on a ground collector plate. The heat energy produced in the nuclear reaction (and which has caused the ionization of ${}^{4}_{2}He$ and ${}^{1}_{1}H$) is transferred in the form of electricity to an external electric circuit.

Fears have been expressed about pollution from such fusion reactors. Given proper technology, these are groundless. The principal reaction products are neutrons, helium and hydrogen nuclei, and radioactive tritium. The latter emits β particles, but it will be recycled, not emitted to the atmosphere.

(ii) Another form of energy, almost infinitely available, is *solar energy*, which reaches the earth's surface at the rate of about 1 kilowatt per square metre.

At present, the materials used to convert solar energy directly are expensive. However, the transistor, which involves the same kind of materials as do solar conversion cells, was reduced in cost by 2 to 3 orders of magnitude by the development of automated manufacturing techniques. A similar fall in cost may occur in the case of solar cells, or newer and cheaper materials may be found in which the conversion occurs. Surprisingly little research is going on in this field.

(iii) The surface of the earth can be regarded as the outside skin of a container of hot liquids. Where hot zones reach the earth's surface, *geothermal power plants* can be set up, and some are now in operation in New Zealand and in Italy. Methods whereby the heat deep within the liquid core of the earth might be used require much research, which so far has hardly begun.

Thus, very large resources of *clean* energy could be available. However, even the building of nuclear fission reactors is going on too slowly to meet the deadline at which we shall need new energy sources.

It is not practicable to turn down a nuclear reactor at night, when the demand for electrical energy drops. *Storage* of the electrical energy constantly being produced by a reactor is an important matter which influences cost. Moreover, we cannot put the extremely heavy atomic reactors into ordinary vehicles. We must have a suitable way of storing electrical energy so that it becomes available on turning a switch.

1.7. *A.D.* 2000 *: the crucial date*

We have not yet put a date on the exhaustion of fossil fuels, the time at which pollution resulting from burning them will become intolerable, or the time when vital materials will be exhausted and have to be recycled. No single answer is possible. However, recent studies imply that by about 1985 the rate of increase of air pollution from cars will rise again after one to two decades of decrease due to exhaust controls. The expected increase in the *number* of cars in the world will outweigh the decreasing emission per car. By about this time, in the U.S.A., most of the oil and natural gas would have to be imported. If our present fossil-fuel based economy is continued, affluent life in the modern sense may not be possible on this planet beyond about 2050. Thus, A.D. 2000 represents a reasonable target date by which new methods of obtaining energy, and of recycling many of our materials, ought to be practically realized.†

Nor is the choice that between an affluent society and the return to a simple agrarian economy. The population of the world is too great for a primitive economy to support. If we had to suffer a sudden substantial decline in our energy *per capita*—and all that it means in, for example, the production and delivery of food—there would be a breakdown of urban life accompanied by starvation and disease.

A viable future is possible. Its basis would be clean and plentiful energy from one or more of the abundant sources named above. But the use of one of the clean sources of energy brings with it an incontestable conclusion : *the form of the new energy will be electrical.* We shall no longer use the solar energy stored in oil or coal or natural gas. Electrochemical science provides the link between electrical and the chemical energy and metallurgical processes on which technology rests. It is as important as a basis of a viable future as are practical new means for attaining a clean and inexhaustible supply of energy.

1.8. *What is electrochemical science ?*

Electrochemistry is the science which deals with the consequences of the transfer of electric charges from one phase to another. When we bring two phases (a metal and a solution containing ions, say) into contact, some of the particles in the original phases tend to transfer to the other phase, carrying their electrical charges (all of the same sign) with them. So one phase has an excess of positive charge and the other an excess of negative charge. There is a potential difference across the region between two phases ; we call this region an interphase.

Electrochemical science deals mainly with *surface situations*, particularly those which are the result of the electrical properties of the interphase.

The electrochemical substance producer (*driven cell*) (*Fig.* 1.5). An electrical energy source maintains a continuous flow of electrons in the metal forming the external part of the circuit and a two-way ionic flow in the electrolyte. At one electrode–electrolyte interface, electrons leave the electrode (*electron source*, *cathode*) and the particles of the solution are electronated, i.e., reduced. At the second interface, the electrode (*electron sink*, *anode*) takes electrons from particles, which therefore are de-electronated, i.e., oxidized. In such a device, the

† The lead-time needed to develop a new energy system might well be some 15 years, and the time to build it perhaps another 15 years.

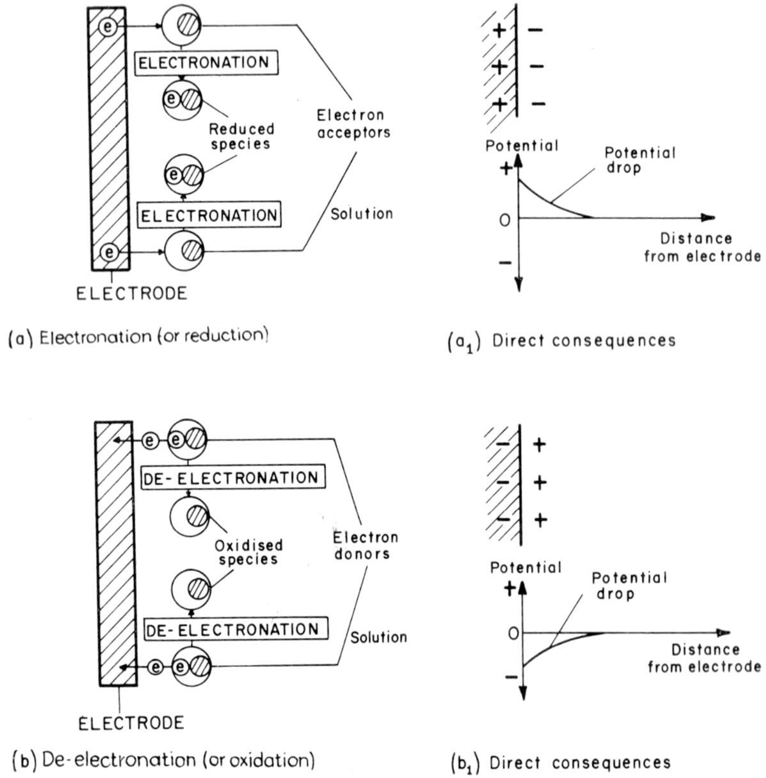

Fig. 1.4. (*a*) In an electronation reaction, a particle on the solution side of the interface accepts an electron from the electrode (reduction). (*b*) In a de-electronation reaction, a particle donates an electron to the electrode (oxidation) (a_1) and (b_1). The most direct consequences are the electrical charging of interphases and the establishing of a potential difference across them.

external energy supply drives the cell. By controlling the rate of electron transfer across the interfaces, and the energy of the electrons, one can control chemical transformations, each of which takes place only when electrons of a specific energy are available. A practical example is the electrolysis of water. In acid solutions, hydrogen ions are electronated to produce hydrogen molecules, and water molecules are de-electronated to produce oxygen molecules :

$$\begin{array}{ll} \text{Cathode} & H^+ + e \rightarrow \tfrac{1}{2}H_2 \\ \text{Anode} & \tfrac{1}{2}H_2O \rightarrow \tfrac{1}{4}O_2 + H^+ + e \\ \hline \text{Overall Reaction} & \tfrac{1}{2}H_2O \rightarrow \tfrac{1}{2}H_2 + \tfrac{1}{4}O_2 \end{array}$$

This device serves to produce hydrogen and oxygen from water.

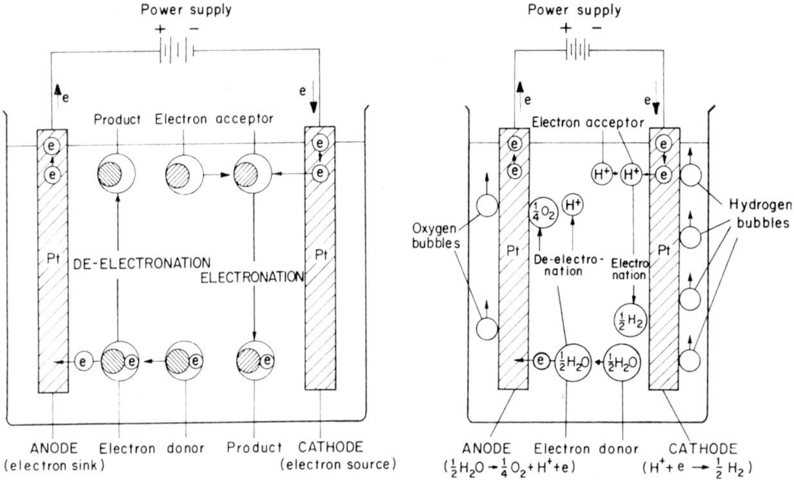

Fig. 1.5. *Left*: the principle of an electrochemical substance producer (driven cell). *Right*: the electrolysis of water in acid solution, by means of a power supply connected to two electrodes immersed in solution.

The electrochemical energy producer (*self-driving cell*). In such a device (fig. 1.6), the electron transfer reactions run spontaneously, rather than by being driven from an external source. Substances are transformed and electrons flow round the circuit. Thus, the electrochemical energy producer generates electrical energy directly. The energy set free in a spontaneous chemical reaction is not experienced as heat. A classical example is the Daniell cell (fig. 1.7).

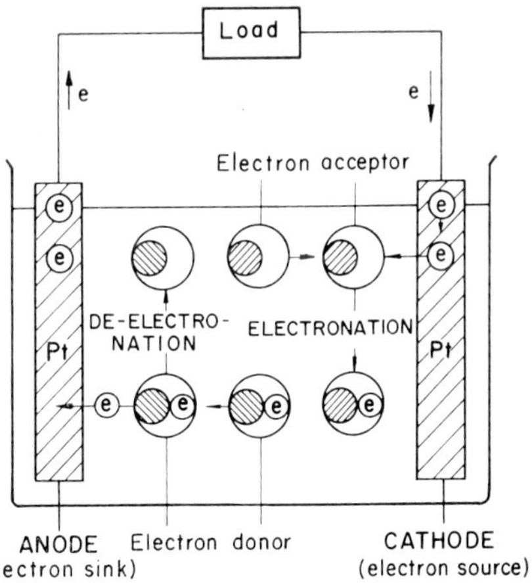

Fig. 1.6. The principle of an electrochemical energy producer (self-driving cell).

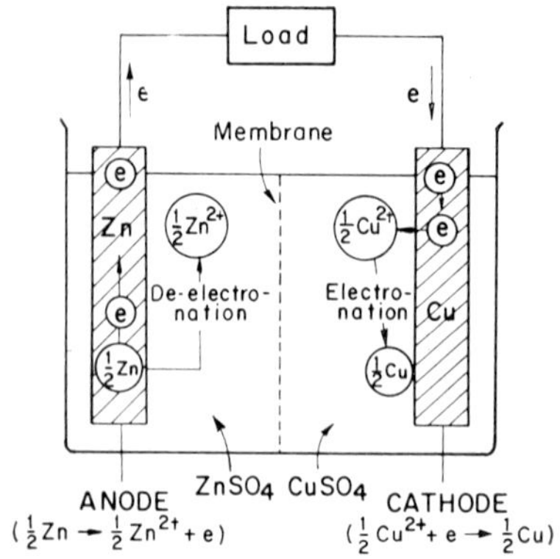

Fig. 1.7. The Daniell cell as a classical example.

A modern example (fig. 1.8) is the hydrogen–oxygen fuel cell, used in space vehicles and discussed in Chapter 7. Basically, the reactions are:

Cathode	$\frac{1}{4}O_2 + H^+ + e \rightarrow \frac{1}{2}H_2O$
Anode	$\frac{1}{2}H_2 \rightarrow H^+ + e$
Overall Reaction	$\frac{1}{2}H_2 + \frac{1}{4}O_2 \rightarrow \frac{1}{2}H_2O$

The overall reaction (occurring *spontaneously*) is the reverse of the overall reaction which is *caused* to take place in the electrolysis of water, by energy supplied from an outside source.

The substance destroyer and energy waster. Let us picture the case of spontaneous electronation and de-electronation reactions occurring on the same piece of metal, so that some areas of the metal act as electron sources, whereas others act as electron sinks. The energy produced is wasted, because—the two interfaces being short-circuited—there is no

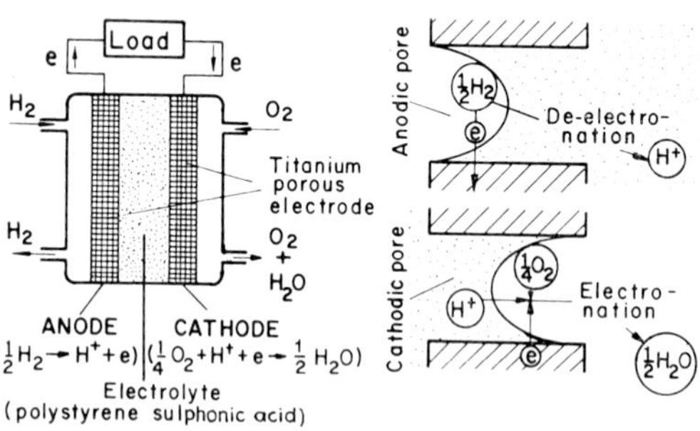

Fig. 1.8. The hydrogen–oxygen fuel cell, and the charge transfer reactions which take place within the pores of the three-phase boundary.

possibility of sending an electric current through an external circuit, as is possible when the two electrodes are separated. *There is also a wastage of substance.* Consider a piece of zinc immersed in dilute acid (fig. 1.9). At the de-electronation areas solution of zinc ($Zn \rightarrow Zn^{2+} + 2e$) is occurring all the time ; this is *corrosion.* By this corrosion, a metal, won from its ores with expenditure of much energy, is converted to ions in solution ; and often, after removal of the water, finishes as an oxide once more.

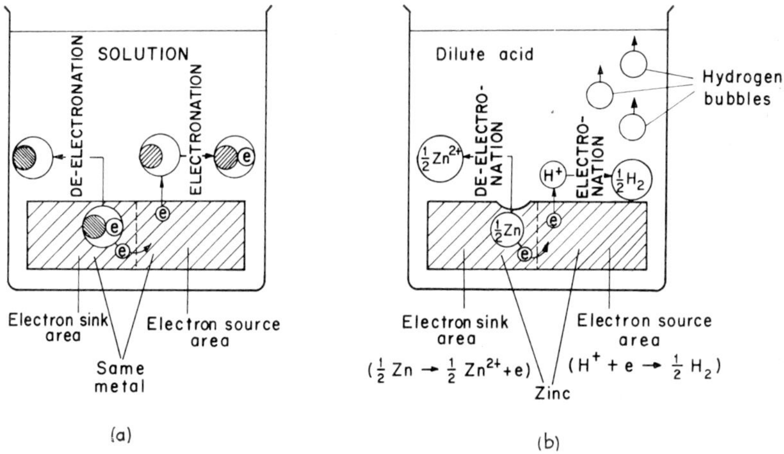

Fig. 1.9. (*a*) Processes occurring in a substance destroyer and energy waster 'undevice'. (*b*) Acid corrosion of zinc is an example of such an undevice.

During the next few decades we shall have to devise a mode of recycling our materials instead of discarding them after use. In the post-fossil fuel age, electrochemical reactions will play a leading role. They will underlie the practical applications of nuclear energy, and for a few decades may be the means of converting chemical energy directly to electrical energy in vehicles. They will provide means of chemical synthesis, of recycling, of the stabilization of materials, of the purification of liquids. They may provide non-polluting alternatives to many chemical processes from which it is difficult to avoid pollution.

Electrochemical science also provides the basis for future developments in biology, for electrochemical processes are involved in the functioning of the nervous system, in blood clotting, in bone growth, and in the functioning of the brain.

CHAPTER 2

ions in water

2.1. *Ionic solutions and electrochemical science*

IONIC solutions are the media in which ions exist; and ions are the reservoir of charges for electrodes, and thus for processes in which substances directly produce electricity and those in which the passage of electricity across electrodes produces new materials.

2.2. *The dissolution process*

2.2.1. *Energy changes during dissolution*

What happens when a salt forms ions in solution? The ions exist already in the crystal lattice of the salt. Each of the two ions has a binding energy in the lattice which is in the region of 200 kJ mol^{-1} for most ions, say 400 kJ mol^{-1} for a two-ion salt (compared with the energy they have as individual ions, separated, and in the gas phase). The heat given up or taken in during solution of 1 mole of a salt is found *experimentally* to be much less than 200 kJ mol^{-1}—just a few kJ per mole of salt. Now, when a substance dissolves, the heat of solution must be the difference between the binding energy of the ions in the crystal lattice, and that of the ions in water. As the heat of solution is so small, the solvation energy must nearly equal the lattice energy, about 200 kJ mol^{-1}. The first quantitative discussion of ion-solvent interactions—which give rise to this solvation energy in the form of heat—was given by Born.

In Born's model, the ion is considered to be equivalent to a charged sphere, the solvent is imagined as a continuum medium with permittivity ϵ, and the ion–solvent interactions are assumed to be entirely *electrostatic* in origin. With these assumptions, the work of transferring an ion from a vacuum into a solvent (the free energy of ion–solvent interactions, $\Delta G_{\text{I-S}}$) will be the difference between the electrostatic energy of the ion in the solvent and that of the ion in a vacuum (see fig. 2.1).

In electrostatics, the energy of a charged conductor carrying a charge $\underset{\sim}{Q}$ at potential V is $\frac{1}{2}\underset{\sim}{Q}V$.

So consider a sphere of radius r, in a vacuum, carrying charge $\underset{\sim}{Q}$. The electrostatic potential at the surface is $\underset{\sim}{Q}/4\pi\epsilon_0 r$, where ϵ_0 is the permittivity of a vacuum, and thus:

$$\text{Electrostatic energy} = \frac{1}{8\pi\epsilon_0}\frac{\underset{\sim}{Q}^2}{r}. \tag{2.1}$$

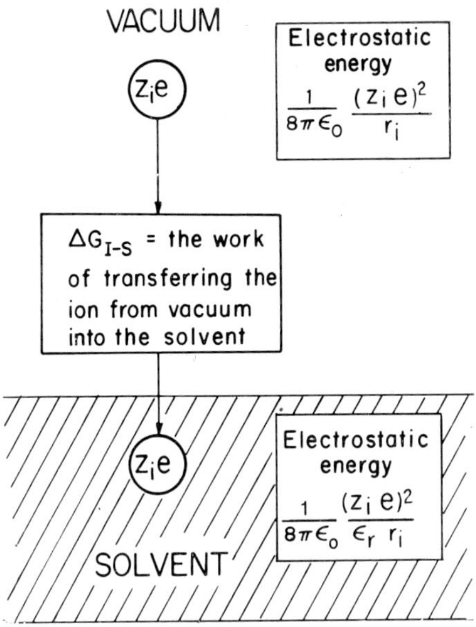

Fig. 2.1. The Born approach to the free energy of the ion–solvent interactions $\Delta G_{I\text{-}S}$.

If we now assume that we can treat an ion as if it were a charged sphere, then the energy of an ion of charge z_ie and radius r_i in vacuum is :

$$\frac{1}{8\pi\epsilon_0}\frac{(z_ie)^2}{r_i}. \tag{2.2}$$

The energy of an ion in the solvent is hence :

$$\frac{1}{8\pi\epsilon_0}\frac{(z_ie)^2}{\epsilon_r r_i} \tag{2.3}$$

where ϵ_r is the relative permittivity of the solvent.

Hence, the energy *change* on transferring the ion from a vacuum to the solvent is :

$$\Delta G_{I\text{-}S} = -\frac{(z_ie)^2}{8\pi\epsilon_0 r_i}\left(1-\frac{1}{\epsilon_r}\right). \tag{2.4}$$

$\Delta G_{I\text{-}S}$ is negative, i.e. the ion is energetically more stable in the solution than in the gas phase.

It is the *heat* of the ion–solvent interactions, $\Delta H_{I\text{-}S}$, rather than the free energy change, $\Delta G_{I\text{-}S}$, which is *experimentally* obtained by measurements with a calorimeter.

$\Delta H_{\text{I-S}}$ can be related to $\Delta G_{\text{I-S}}$ by the thermodynamic relation :

$$\Delta H_{\text{I-S}} = \Delta G_{\text{I-S}} - T\left(\frac{\partial \Delta G_{\text{I-S}}}{\partial T}\right)_p \qquad (2.5)$$

where the derivative in the right-hand side is taken at constant pressure.

Now, if one differentiates the above expression (2.4) :

$$\left(\frac{\partial \Delta G_{\text{I-S}}}{\partial T}\right)_p = -\frac{(z_i e)^2}{8\pi\epsilon_0 r_i}\frac{1}{\epsilon_r^2}\frac{\partial \epsilon_r}{\partial T}$$

and thus :

$$\Delta H_{\text{I-S}} = -N_A \frac{(z_i e)^2}{8\pi\epsilon_0 r_i}\left(1 - \frac{1}{\epsilon_r} - \frac{T}{\epsilon_r^2}\frac{\partial \epsilon_r}{\partial T}\right) \text{ per mole of ions.} \qquad (2.6)$$

When one compares the calculations from Born's equation (2.6) with the results obtained experimentally, one finds a large discrepancy between the two sets of values. The Born values are in some cases 50 per cent higher.

There must, therefore, be another process which we have to take into account ; it concerns the special structure of water molecules which forms around an ion, differing from the structure of water in the solvent. When we argue that $\Delta H_{\text{I-S}}$ is given by (2.6), we are talking about a solvent which is an idealized continuum, without structural features appearing locally near the ion (such as result with actual water from the interaction between the ion's electric field and nearby water dipoles). For the continuum, we can use a simple dielectric constant, correct for the interior of water and independent of the influence of ions. But when water molecules get near an ion, the ion's electric field breaks up the old structure, their orientation with respect to each other is altered and the energy of interaction of the solvent with the ion differs greatly from the idealized continuum picture. We require *a treatment of the ion–solvent interactions which takes this into account.*

The water molecule can be regarded roughly as an electric dipole, i.e. it can be considered as two equal and opposite charges $(-q, +q)$, separated by a small distance l. The product $\mu_W = ql$ is termed the *dipole moment* of the water molecule.

In a homogeneous electric field the dipole is aligned along the direction of the field, which is ordinary dielectric polarization. In a non-homogeneous electric field, the dipole not only aligns itself along the field, but also gets pulled in the direction of increasing electric field strength. The electric field strength due to an ion in its neighbourhood increases as one approaches the ion, so the water dipole gets pulled towards the ion, the pull increasing as the distance between the two decreases.

Because of these ion–dipole forces, each ion is enveloped by a sheath of immobilized water dipoles (the *primary region*), and only far away from the ion is the structure of water undisturbed (the *bulk region*). But the primary hydrated ion does cause some disturbance of the surrounding water; there is a *secondary region*, between the primary region and the bulk region, in which the water molecules are partly oriented. These three regions are shown in fig. 2.2.

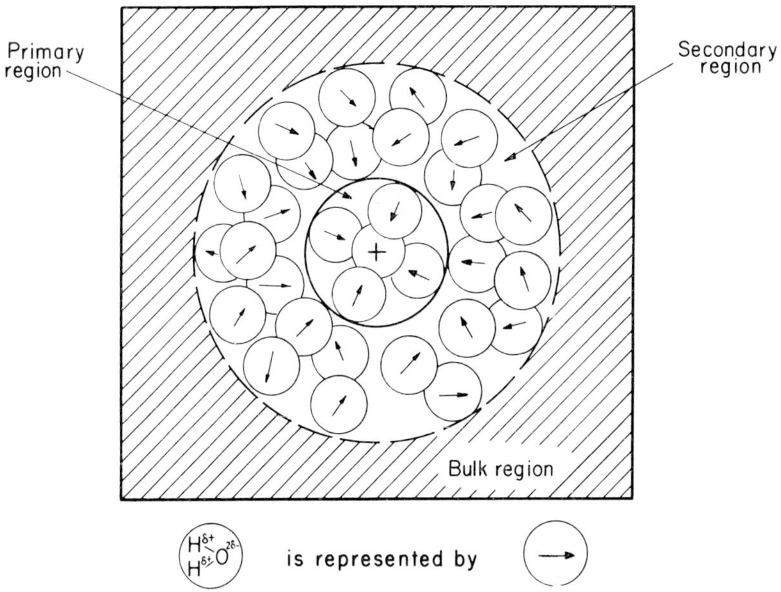

Fig. 2.2. The structure of water near an ion according to Bernal and Fowler (the ion is surrounded by four completely oriented water molecules in the primary region).

Thus, there are three principal contributions to the solvation energy:

A Born term for the interaction of the ion with the almost undisturbed region of solvent, far from the ion;

The ion–dipole interactions ($\Delta H_{\text{I-D}}$) for the region close to the ion;

And the energy change due to the re-orientation of water molecules in the secondary region (fig. 2.2) which leads to the breakage of some hydrogen bonds and the formation of others.

Let us call the latter contribution $\Delta H_{\text{struct. break}}$. Thus:

$$\Delta H_{\text{I-S}} = -N_{\text{A}} \frac{(z_{\text{i}}e)^2}{8\pi\epsilon_0(r_{\text{i}}+2r_{\text{w}})} \times \left(1-\frac{1}{\epsilon_r}-\frac{T}{\epsilon_r^2}\frac{\partial\epsilon_r}{\partial T}\right)+\Delta H_{\text{I-D}}+\Delta H_{\text{struct. break}} \quad (2.7)$$

where the Born charging contribution now refers to an ion of radius $r_i + 2r_w$ (r_w = the radius of the water molecule). In this way, one assumes that the region of solvent outside the first water layer round the ion is effectively a continuous dielectric ; for here the effect of the ion on the water is weak enough to allow it to retain (almost) the permittivity of the undisturbed solvent.

The electrostatic force P between an ion and a dipole (fig. 2.3) is the difference between the two forces P_- and P_+ shown, and $(P_- - P_+)$ works out approximately to

$$P = \frac{z_i e \mu_w}{2\pi\epsilon_0 r^3}. \tag{2.8}$$

The work involved in bringing a water dipole from infinity to the surface of the ion is thus :

$$\int_{\infty}^{r_i + r_w} P \, dr = -\frac{z_i e \mu_w}{4\pi\epsilon_0 (r_i + r_w)^2}. \tag{2.9}$$

If we now suppose (to simplify the argument) that one ion is surrounded just by 4 oriented water molecules, the expression $\Delta H_{I\text{-}D}$, per mole of ions, can be taken as :

$$\Delta H_{I\text{-}D} = -N_A \frac{z_i e \mu_w}{\pi\epsilon_0 (r_i + r_w)^2}. \tag{2.10}$$

The structural treatment just outlined is known as the ion–dipole approach to the heat of solvation, and it was first carried out by Bernal and Fowler (1933). Their final equation is :

$$\Delta H_{I\text{-}S} = \Delta H_{\text{struct. break}} - N_A \frac{z_i e \mu_w}{\pi\epsilon_0 (r_i + r_w)^2} - N_A \frac{(z_i e)^2}{8\pi\epsilon_0 (r_i + 2r_w)^2}\left(1 - \frac{1}{\epsilon_r} - \frac{T}{\epsilon_r^2}\frac{\partial \epsilon_r}{\partial T}\right). \tag{2.11}$$

The term $\Delta H_{\text{struct. break}}$ in (2.7) has a numerical value of about 80 kJ mol^{-1} for water molecules near to cations such as Na^+ and almost 120 kJ mol^{-1} for those near anions such as Cl^-. The value of $\Delta H_{\text{struct. break}}$ differs for cations and anions because the orientation of water molecules in the first layer, and therefore the amount of hydrogen bond breaking, depends on whether the ion is a cation (positively charged) or an anion (negatively charged).

The calculation of solvation energy in this way is still crude ; there are several considerations which we have neglected in this simple approach. For example, it has been more recently realized that the interaction between each water molecule and the ion is better represented if the quadrupole structure of water (a partial positive charge

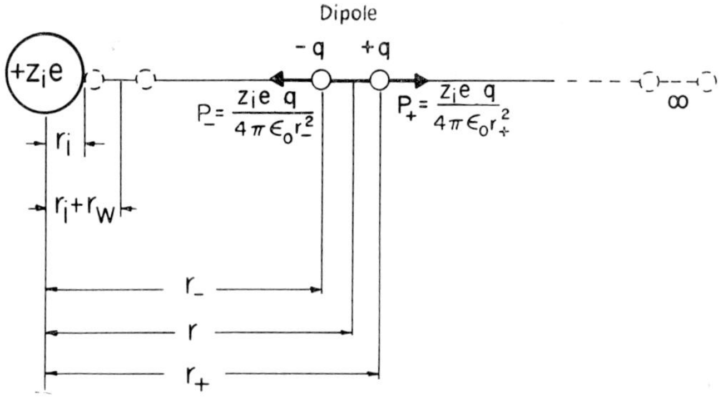

Figure 2.3. The electrostatic force between an ion and a dipole (in vacuum).

on each hydrogen, balanced by negative charges on oxygen) is taken into account. On this basis, an ion–quadrupole calculation of solvation energy has been made. It is outside the scope of this book. Suffice it to say that the calculated $\Delta H_{I\text{-}S}$ values are in better agreement with experimental data than are those based on the earlier view.

A comparison of the values of solvation energies from the dipole and quadrupole approaches is shown in table 2.1.

2.3. *Structure of the region near the ion*

Our interpretation of solvation energy is approximate, insofar as we have considered a region of special *structure* near the ions and have not said of what it consists, except the vague description 'oriented dipoles' (more correctly, quadrupoles).

Ion	Ion-dipole approach, eqn. (2.11) kJ mol^{-1}	Ion-quadrupole approach kJ mol^{-1}	Experimental values kJ mol^{-1}
Li^+	−670	−640	−521
Na^+	−499	−457	−406
K^+	−379	−346	−326
Rb^+	−343	−307	−301
Cs^+	−300	−266	−277
F^-	−329	−508	−505
Cl^-	−238	−346	−363
Br^-	−216	−310	−336
I^-	−187	−265	−295

Table 2.1. Comparison of experimental values of $\Delta H_{I\text{-}S}$ with those predicted by the ion-dipole and ion-quadrupole approaches.

We can find out more about this structure from a knowledge of the *compressibility* of solutions. We can find the compressibility (β) of pure water, and of a solution of, say, 1 molar sodium chloride. The experimental results show that the β of a solution is less than that of water. We argue that this is so because the water quadrupoles in a region round the ion, the 'solvation shell', are already compressed so much by the electrical field of the ion acting upon them that they cannot be compressed further (fig. 2.4).

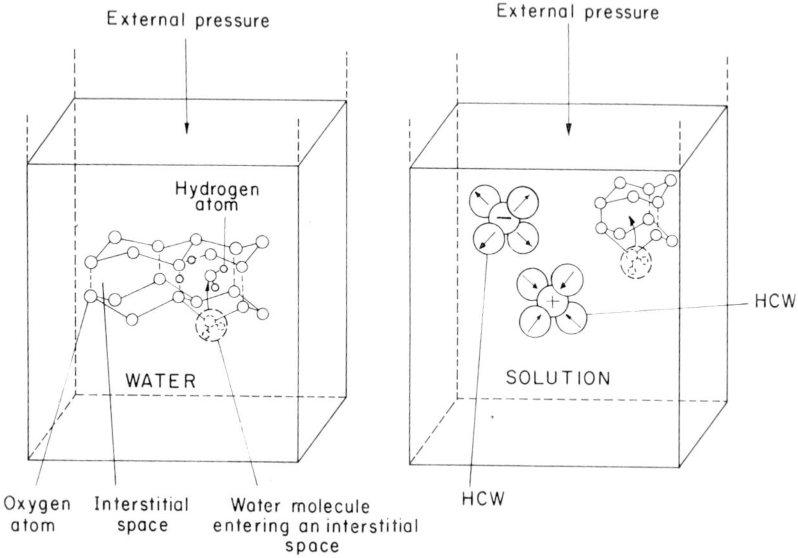

Fig. 2.4. *Left*: under external pressure, a water molecule breaks its hydrogen bonds with the network and occupies an interstitial space. *Right*: the hydration shell is compressed by the field of the ion, and cannot be compressed further by the external pressure; HCW = hydration-coordination water molecule. This is a water molecule which remains associated with the ion when it moves through the solution. The arrows indicate the dipole moment of the water molecules.

Hence, the compressibility of the *solution* (compared with that of water) is diminished by an amount which depends upon how many water molecules each ion is firmly attached to. By measuring the compressibility of the solution, and by calculating how much the compressibility of the solution is decreased by one water molecule being attached to the ion, we can work back from a knowledge of the ratio $\beta_{\text{soln}}/\beta_{\text{water}}$ to the total number of water molecules tightly attached to all the ions in the dissolved salt; that is to the number of moles of water molecules per mole of salt, which is the 'solvation number' of the salt.

However, we still need to know the actual number of water molecules around *each* ion. Hence, we want to distinguish between the number attached to a cation and the number attached to an anion. One method which allows this is the measurement of what are called ' ionic vibration potentials '. Ultrasonic stationary waves generated within an electrolyte cause both cations and anions to oscillate at an antinode, at the frequency of the sound. The amplitude of oscillation depends on the inertia (that is, the mass) of an ion ; the greater the mass, the smaller the amplitude. In an undisturbed ionic solution there is, on a time average, no separation between cations and anions ; but separation occurs at the antinodes of the stationary wave pattern in an ionic solution. Suppose that the heavier ones (including the solvation sheath†) are the cations, for example, $Li^{+}(n\ H_2O)$, and the lighter ones the anions, such as $Br^{-}(m\ H_2O)$. If we insert two electrodes close to antinodes at *about* the same distance apart as half the wavelength of the sound (fig. 2.5), we should be able to pick up an electrical potential difference due to this separation. In effect, the cations and the anions then form a series of separated layers (like a series of double layers, see Chapter 3) in the solution.

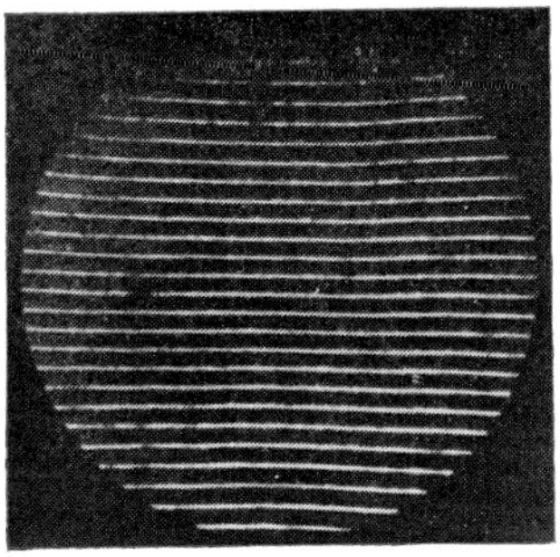

Fig. 2.5. An ultrasonic standing wave system in a liquid, made visible by optical means and then photographed.

An expression which relates this ionic vibration potential to quantities connected with the mass of each ion–solvent entity was calculated by Debye, but the argument is too difficult to discuss here. The important point is that we get from it the *difference* of the mass of the cation including its *attached water molecules*, and the mass of the anion with *its attached water molecules*. We know the masses of the ions ; so we can obtain the difference in the masses (and hence in the numbers) of water molecules which remain attached to each during its oscillation.

† Defined, then, as those water molecules which stay with the ion during the oscillation caused by the sound waves.

The measurements of compressibility outlined above gave the *sum* of the numbers of water molecules attached to the ions. Hence, from the two types of measurements, one knows both the sum of and the difference between the numbers of water molecules attached to anion and cation. So each of these numbers can be evaluated.

Table 2.2 gives some values for the ' hydration number ', by which is meant the number (per ion) of water molecules which remain with the ion for a significant time during its thermal movements in the solution.

Ion	From compressibilities and ionic vibration potential measurements	Average from other methods
Li^+	4·5	5 ± 1
Na^+	4·5	5 ± 1
K^+	3·8	4 ± 2
Rb^+	3·0	3 ± 1
Cs^+	2·5	—
F^-	4·0	4 ± 1
Cl^-	2·2	1 ± 1
Br^-	1·8	1 ± 1
I^-	1·5	1 ± 1

Table 2.2. Hydration numbers.

How one proceeds from information about these hydration numbers to the detailed structure of water molecules around an ion is a longer story. It includes trials of various imagined arrangements of water molecules near ions, and comparison of the heats of solvation calculated assuming each model arrangement with the value obtained experimentally.

The structure which gives the best fit between model trials and experimental values is shown in fig. 2.6.

2.3.1. *What use is the study of ion-solvent interaction to electrochemical science?*

Knowledge of the energy of ions in solution is important in electrochemical science. For example, the depth of the energy well occupied by an ion in solution determines the difficulty of transferring an electron to the ion. It is easy to understand that, if an ion is tightly bonded to the solvent, it will not be so easy for an electron to neutralize it as when it is less tightly bound. Later on (Chapter 4) we shall find that the solvation energy influences a quantity called ' overpotential ', which determines the potential difference which has to be applied to a cell to make an electrochemical reaction occur in it. Indirectly, the

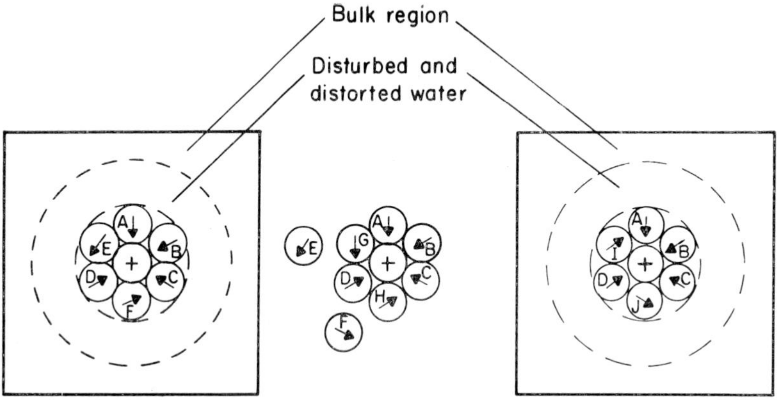

Fig. 2.6. The structure of the region near an ion (Bockris and Reddy model). A, B, C, D = Hydration-coordination water molecules (such water molecules join the ion in its movement through the solution). E, F, G, H, I, J are non-hydration-coordination water molecules, which are not firmly oriented and are associated with the ion only in its momentary stationary positions; they are, therefore, left behind when the ion moves (see the water molecules E and F).

degree of solvation influences the energy (and hence the cost) of making some electrochemical processes work, since the energy is (*p.d.* × *current* × *time*).

Ion–solvent interaction also affects the solubility of an ionic substance: the greater the interaction, the greater is its influence on the solubility of salts in non-aqueous solutions. For example, whether sodium acetate will be soluble in acetamide depends on the heat of solvation of sodium ions and acetate ions in acetamide. The solubility can be predicted approximately from the energy of the ion–solvent interaction.

2.4. *Ion–ion interactions in solution*

2.4.1. *Properties of solutions depend on the concentration of the substance dissolved*

We have discussed the energy of an ion with respect to its electrical interactions with the solvent. There is a more subtle energy of interaction which determines the properties in solution, that between ion and ion. As the concentration of a solution rises (and therefore the average distance between particles decreases), so the energy of interaction between cations and anions increases, as a consequence of Coulomb's law.

It follows, qualitatively, that the properties of ionic solutions (essentially dependent upon the mean energy of the ions) will depend upon the concentration of the ions.

2.4.2. *Ideal and non-ideal solutions*

The terms ideal and non-ideal arise from the theory of gases. It is possible to show that the free energy of a mole of gas depends upon pressure in the following way :

$$\mu_i = \mu_{0i} + RT \ln p_i \tag{2.12}$$

where p_i is the partial pressure of the component i in the gas mixture, and μ_{0i} is the *standard* free energy, i.e. the free energy when the partial pressure p_i is unity. It is called the *chemical potential* of i.

However, equation (2.12) is only true if there is negligible interaction energy between particles making up the system (true for gas molecules, because they do not carry charges, and hence have zero coulombic interaction according to Coulomb's law). In, for example, a solution of Li^+ and Br^- ions, there are strong interactions between the positive and negative ions. To allow for this, we introduce another term into equation (2.12) (which was only correct for gases at fairly low pressures), and write this other term as :

$$RT \ln f_i.$$

Hence :

$$\mu_i = \mu_{0i} + RT \ln f_i c_i = \mu_{0i} + RT \ln a_i \tag{2.13}$$

where a_i is called the activity of the ionic species, i, and we have used c_i for the concentration of an ion in solution in place of p_i, the pressure in the gas.

Here, the activity of an ionic species, i, is the product of the concentration of the ion in the solution and a term f_i ; $RT \ln f_i$ is the ionic attraction energy between a mole of positive and a mole of negative ions, at a concentration c.

When the solution is very dilute, the ions will be so far apart that the energy of their interactions will be negligible. If this is so, then the ' energy of unideality ' (which is what we call the ions' energy of interaction with each other)† must tend to zero. Hence, $RT \ln f_i$ must tend to zero, and it then follows that f_i tends to 1 (and $\ln f_i$ to 0) as the concentration of ions in the solution diminishes. An *ideal* solution (no interaction energy), therefore, has an activity coefficient of 1.

Corresponding to this concept of the activity coefficient, f_+ or f_-, of an individual cationic or anionic species respectively, there is another kind of activity coefficient, that for a given salt in an electrolyte. It

† It is the energy which makes an ' ideal ' expression (2.12 with c_i instead of p_i) no longer applicable.

is easy to show that this *mean activity coefficient* of the substance is given by :

$$f_{\pm}=(f_{+}f_{-})^{1/2} \quad (2.14)$$

if the salt is 1 : 1 (i.e. like NaCl) and not 1 : 2 (like $BaCl_2$), etc.

As $c \rightarrow 0$, $f_{\pm} \rightarrow 1$. Thus, as HCl is diluted from 0·10 to 0·0001 molal, $f_{\pm}$ rises from 0·794 to 0·989. (See Table 2.3).

Concentration (molality)	Mean activity coefficient
0·10	0·794
0·05	0·830
0·02	0·876
0·01	0·905
0·005	0·929
0·002	0·952
0·001	0·966
0·0005	0·975
0·0002	0·984
0·0001	0·989

Molality = moles per kilogram of solvent.

Table 2.3. Mean activity coefficient of HCl at different concentrations at 298 K.

2.4.3. *The theory of interionic interaction*

There is an accepted theory for interionic interaction (at least in dilute solutions).

We first imagine a cloud of charges surrounding the single ion we choose to consider. Let us compare the situation to a dance, at which about half of the dancers are female, and half male. We assume that each man wishes to dance with all the ladies and each lady wishes to dance with all the men, and either side can take the initiative and ask. There will be a large number of encounters and some will result only in brief partnerships. And, at any one instant, there will be a large number of people who have not got partners, because they are in the process of changing them, or for other reasons.

However, if we look down at the dance floor from above, and take many thousands of instantaneous photographs of it, we shall find, taking the average of all the situations indicated by the many different photographs, that each man tends to be surrounded by more women than men, and each woman tends to be nearer to a man than she is to another woman.

The situation in solution is similar. If we consider a positive ion, say, it will have in its neighbourhood more anions than cations. The *net* charge on the ions surrounding a positively charged ion (cation) will be negative. All the positive and negative ions which surround a given positive ion are jiggling around and about in solution, so that they can be regarded as *smoothed* out. Consequently, it is reasonable to speak of a ' charge cloud ' surrounding each ion ; see fig. 2.7. Ions of both signs of charge inhabit the cloud, but those with a charge opposite to the central ion predominate.

What applies to the cation also applies to the anion. Each anion has, on an average, a greater number of cations than other anions surrounding it.

The net charge on the surrounding cloud is equal and opposite to that of the central ion. How is this net charge distributed in space with respect to the distance from the central ion we chose to use as a reference point ? Let us divide the space around the central ion into spherical shells of constant thickness, dr. It can be shown that the net charge, dq, on such a spherical shell varies with the radius of the sphere, and has a maximum value for a radius called ' the Debye–Hückel length ' or ' the ionic-cloud thickness ' (see fig. 2.7). The symbol κ is used to represent the reciprocal of this ionic cloud thickness.

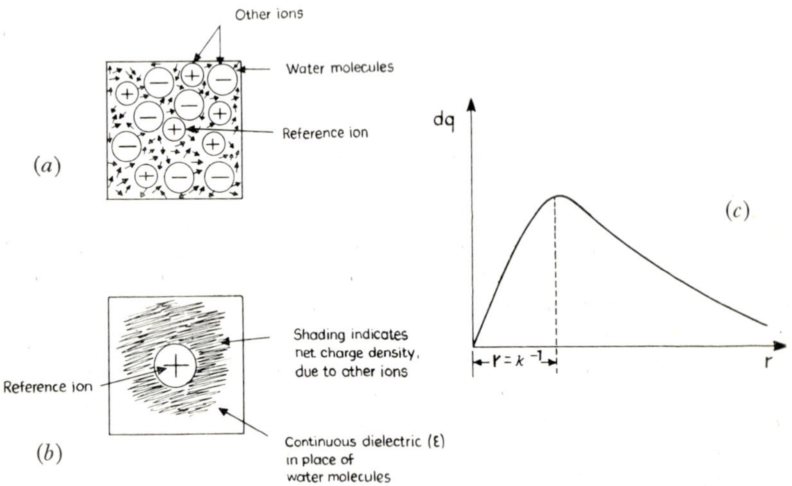

Fig. 2.7. A schematic comparison of (*a*) the assembly of ions and solvent molecules which constitute a real electrolytic solution ; (*b*) the Debye–Hückel picture in which a reference ion is surrounded by net charge density, due to ions of opposite sign and a dielectric continuum of the same dielectric constant ϵ as the bulk solvent. Diagram (*c*) illustrates the net charge, dq, enclosed in spherical shells of the same thickness, dr, but various radii, r, having a maximum value for $r = \kappa^{-1}$.

The theory of the ionic atmosphere was elaborated by Debye and Hückel. For a uni-univalent electrolyte this theory gives :

$$\kappa^{-1} = 1/B\sqrt{c} \quad (2.15)$$

where c is the electrolyte concentration and the value of B varies with the temperature and the relative permittivity.

The ' energy of unideality ' may be calculated if we find what *extra* energy is gained by an ion, due to the existence of this cloud of oppositely charged ions.

The calculation is a famous application of statistical mechanics to solutions, the analogue, for solutions, of Maxwell's calculation of the distribution of velocities in gas molecules. Of course, because the interaction energy between ions in solution is equal to $RT \ln f_{\pm}$, one can also regard it as a calculation of the mean activity coefficient. We here state the result, which is :

$$RT \log f_{\pm} = -\frac{A\sqrt{I}}{1+B\alpha\sqrt{I}} - 2{\cdot}303\ RT\frac{n_{\rm h}}{n}\log a_{\rm w} + 2{\cdot}303\ RT \log\frac{n_{\rm w}+n}{n_{\rm w}+n-n_{\rm h}}. \quad (2.16)$$

For water as the solvent, n is the number of moles of electrolyte present in $n_{\rm w}$ moles of water, $n_{\rm h}$ is the sum of the moles of water in the primary hydration sheaths for both cations and anions, A and B are constants for a given temperature, and $a_{\rm w}$ is the water activity which can be obtained from independent experiments. The quantity I is known as ' the ionic strength ' and is given for one electrolyte in solution, cation charge z_+ and anion charge z_-, as :

$$I = \tfrac{1}{2}(c_+ z_+^2 + c_- z_-^2). \quad (2.16')$$

In a uni-univalent electrolyte, $c_+ = c_- = c$, $z_+^2 = z_-^2 = 1$, and thus $I = c$; the ionic strength and the electrolyte concentration are identical. Finally, α is ' the ion size parameter '. The value of α is not precisely known, but it lies between the sum of the crystallographic radii of the ions and the sum of the radii of the ions with their water sheaths.

In equation (2.16), the first term on the right-hand side takes into account the ion–ion interactions, and the last two terms (due to Stokes and Robinson), the ion–solvent interactions. *Equation* (2.16) *unites these two basic aspects of the interactions* (*ion–ion and ion–solvent*) *in an electrolytic solution.* It agrees fairly well with experiment†.

† The agreement does not mean that research into the theory of solutions is all done ; indeed, the agreement between the theory represented by equation (2.16) and experiment in the intermediate concentration range is fortuitous, because the way in which the ion–ion interactions are calculated is not thought to be correct except at very low concentrations, say below 0·005 M.

2.4.4. *Why activity coefficients?*

We had to consider ionic solvation because it was part of *understanding* the stability of an ion in solution, and how solvation influenced solubility.

We have also to consider the energy which ions get by interacting together in solution. We can't do calculations on electrode processes, and find out for example how to persuade sewage to yield useful energy, unless we know something about the energies of the ions in solution from which we hope that the electrodes will enable us to extract the energy.

Another reason for knowing something about these activity coefficients is that when we want to calculate the potential of an electrode in solution as a function of the concentration of the ions present, in the special case when the exchange of electric charges between the ion and solution has an equal rate in both directions, this electrode potential, V_{rev}, can be shown to be given by (see Chapter 5):

$$V_{rev} = V_{rev}{}^0 + \frac{RT}{F} \ln c_i + \frac{RT}{F} \ln f_i.$$

Thus, we cannot do the most simple calculation in the theory of electrode processes (which always involves V_{rev}), without knowing the activity coefficient of the electrolyte.

2.5. *Incomplete dissociation and complexes in solution*

2.5.1. *Structure in solution*

The above description of a solution deals with the sorting and distribution of ions with respect to each other, how they join up and interact with solvent molecules, and how they interact with each other and affect the energy (and thereby the properties) of the solution. Ions collide in solution, but they rebound, and we have been concerned with the time-average situation among many thousands of millions of collisions.

It could be that ions do not immediately bounce off each other but stay 'stuck' together temporarily ('ionic dipoles'), or, in a more extreme sense, that some cation will bond with one or more anions. Then, *complex ions* (two or more ions associated together for time exceeding, say, 1 s) may exist in solution.

Temporary association into ionic dipoles will mean that a fraction of the ions, at any moment, will not be available to function *as ions*. The associated ion-pairs could be treated as being in equilibrium with the corresponding free ions: $C^+ + A^- \rightleftharpoons C^+A^-$; if the law of mass action is applied to this equilibrium,

$$K_{assoc} = a_{(C^+A^-)}/a_{(C^+)}a_{(A^-)} \qquad (2.17)$$

where $a_{(\ )}$ represents the activity of the corresponding species and is a constant at a given temperature and pressure. K_{assoc} is called 'the association constant', and its value expresses the tendency of the free ions to associate temporarily together into pairs. Table 2.4 shows some values of log K_{assoc}.

	Sodium	Silver	Magnesium	Calcium
Hydroxide	−0·7	2·3	2·58	1·30
Chloride	Nil	3·2	Nil	Nil
Iodate	−0·5	0·6	0·72	0·89
Nitrate	−0·6	−0·2	Nil	0·28
Sulphate	0·7	1·3	2·23	2·28
Thiosulphate	0·6	8·8	1·83	1·95

Nil indicates that ion-pair formation is negligible.

Table 2.4. Log K_{assoc} values of ion-pairs in aqueous solutions (298 K). Data from : C. W. Davies, *Ion-Association,* Butterworths, London 1962.

If the activity of an ion is very low compared with what would be expected from the product of its concentration and the activity coefficient, calculated assuming only ionic attraction between nearby ions, then it is reasonable to conclude that a longer-time bonding is occurring. However, such an inferential method is not always satisfactory because it may be difficult to calculate activity coefficients from theory alone, particularly for higher valency electrolytes, or at concentrations above very low ones (e.g. 0·01 M).

Studies of absorption spectra are useful here. The constitution of a chemical entity, the bonds it forms, the strength of the attachment of the atoms together, the bending of bonds, their twisting, all give rise to energy levels characteristic of the entity, and associated each with a definite energy. Thus, if we measure the intensity of radiation (using a wide band of frequencies) entering a cell containing such molecular entities, and measure the intensity of frequencies over the whole range when it comes out, we shall observe a decrease in intensity of the transmitted radiation just at those frequencies at which it has been absorbed by the substance in solution. This fundamental idea is associated with the effect known as *resonance.* If the energy of the photons of light of a particular wavelength matches the difference between a pair of energy levels, then photons are absorbed. Therefore, in an *absorption spectrum* (a graph of transmitted intensity against wavelength or frequency), the minima will show us the wavelengths or frequencies at which absorption has occurred. If we take various hypotheses about the motions to which these minima correspond, and calculate the frequency associated with various structural models, we can see which model fits the observation. For example, it might be an oscillating bond between two atoms which provides the resonance absorption, or rotation of the molecule. Knowing this gives information on what is happening in solution.

Some examples of complex ions are given in table 2.5.

Ion	Solvent	T/K	Complex	Configuration
Ti^{3+}	LiCl–KCl	673	$TiCl_6^{3-}$	DO
V^{3+}	$CaAlCl_4$	1073	VCl_4^{-}	DT
V^{3+}	LiCl–KCl	673	VCl_6^{3-}	O
V^{2+}	Cs_2ZnCl_4	1173	VCl_4^{2-}	DT
V^{2+}	LiCl–KCl	673	VCl_6^{4-}	O
Cr^{3+}	LiCl–KCl	673–1273	$CrCl_6^{3-}$	O
Cr^{2+}	LiCl–KCl	673–1273	$CrCl_4^{2-}$	DT
Mn^{2+}	LiCl–KCl	673–1273	$MnCl_4^{2-}$	T
Fe^{2+}	LiCl–KCl	1273	$FeCl_4^{2-}$	T
Co^{2+}	LiCl–KCl	673–1273	$CoCl_4^{2-}$	T
Ni^{2+}	LiCl–KCl	973–1273	$NiCl_4^{2-}$	DT
Cu^{2+}	LiCl–KCl	673–1273	$CuCl_4^{2-}$	DT

DO = distorted octahedral; DT = distorted tetrahedral; T = tetrahedral; O = octahedral.

Table 2.5. Configurations of some chlorocomplexes in molten chlorides (data from: G. J. Janz, *Molten Salts Handbook*, Academic Press, New York–London, 1967, p. 368).

2.6. *Transport*

The justification for a chapter on solutions in this book about electrochemical science is that one cannot have things happening at an electrified surface in contact with a solution without a corresponding reservoir of ions in solution. The demand for electric charge at the interface cannot be met by ions which remain stationary in the solution, or just move about at random. There is a *logistic* part in electrochemistry. The ions and the particles have each to move up (at a calculable rate) to the interphase between solution and electrode to receive an electron from the electrode, or to give up an electron to it.

There are three ways in which ions reach the electrode, and then withdraw from it.

(i) Diffusion is the easiest to understand. If an electrode is giving electrons to metal cations, there will be a concentration gradient towards the electrode; the concentration of the metal ions will get less as the electrode is approached, because the electrode is removing a lot from the solution. Diffusion occurs down the concentration gradient towards the electrode. The equation is:

$$i_{\mathrm{diff}} = -DF(\mathrm{d}c/\mathrm{d}x) \qquad (2.19)$$

where i_{diff} is the rate of diffusion of ions to the electrode expressed in $C\,m^{-2}\,s^{-1}$; D is the diffusion coefficient; F is the Faraday constant (=96 500 C); and dc/dx is the concentration gradient near the electrode.

Considering this equation enables one to understand how electrode reactions can occur. Suppose a small current is passing through the electrode. Then the concentration gradient needed to make $-DF(dc/dx)$ equal to this small current (i.e. to make the ions diffuse up at a rate equal to the rate at which their partner electrons are arriving at the surface) will be itself correspondingly small. Thus, the concentration *at* the electrode surface ($dc = c_{surface} - c_{bulk}$) will be scarcely different from that in the bulk.

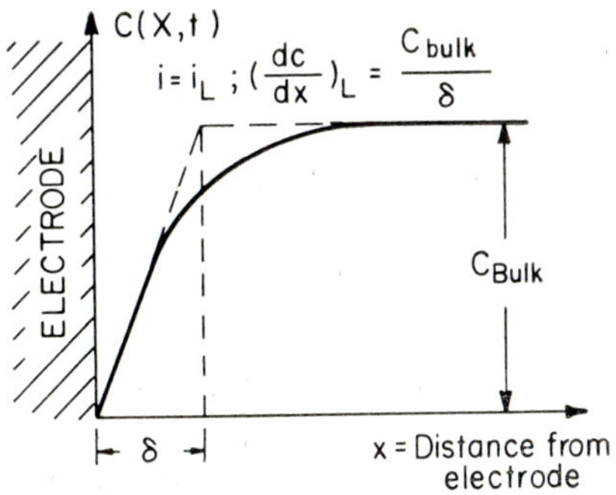

Fig. 2.8. In this figure, two approximations are made: the diffusion occurs across a *Nernst diffusion layer* of constant thickness δ, and the approximation of a linear concentration gradient. As we can see, dc/dx has a maximum value for $c_{electrode} = 0$, and this value corresponds to *the limiting current density*, i_L.

However, as the current density increases dc/dx must increase (cf. 2.19). Does dc/dx have a limiting maximum value? To understand the idea of a maximum value for dc/dx in a simple way, let us assume a linear concentration gradient (see fig. 2.8). We can see that the maximum concentration gradient will be:

$$\frac{dc}{dx} = \frac{c_{bulk} - c_{electrode}}{\delta} = \frac{c_{bulk}}{\delta} \qquad (2.20)$$

and that a *limiting* current (cf. 2.19), the highest which can exist in a given electrolyte for the deposition of a given species, will be observed for $c_{electrode} = 0$.

We shall develop this theme further in Chapter 4.

(ii) Another mode of transport of ions in solution is one in which the electric field produced between the electrodes pulls the ions through the solution towards the electrodes. The field accelerates the ions, pulling the cations towards the negative electrode (cathode) and the anions towards the positive one (anode). Thus, the total rate of arrival of ions at an electrode will depend on the concentration gradient and also on those ions brought up independently of the concentration gradient, as a result of the electric field in the solution.

This latter type of transport is sometimes called *Faradaic*, after Faraday, who first studied the transport of ions to and from electrodes under applied electric fields.

(iii) The third mode of transport of ions in solutions is by convection. There are two types of transport by convection.

—*Forced convection.* In a stirred solution ions are forced to the electrode interface.

—*Natural convection.* During deposition at an electrode, ions are removed from the solution to become atoms on the electrode. The layer of solution near to the electrode becomes depleted of ions, so causing *density* differences and hence convection ' stirring ' in the vicinity of the electrode (see fig. 2.9).

2.7. *Transport numbers*

Apart from the situation very near the electrodes, where diffusion is helping the transport of ions to and from the interfaces, there is an intermediate region, between the electrodes, where only the electric field makes the ions move. This is because far away from the electrodes there are no concentration gradients. Here, a strange thing happens : the ions do not move at equal rates, in fact anions ($-$) often move faster than cations ($+$). If the total current density carried is i, it is divided into two parts, i_+ and i_-. Of course, $i_+ + i_- = i$, the current density observed in the external electrical circuit.

We define a quantity called transport number as :

$$t_+ = i_+/i\,; \quad t_- = i_-/i. \tag{2.21}$$

But how can it be that the ions are going to arrive in unequal numbers at the electrodes ? The ions are just electric charge carriers as far as the electrode is concerned, and $i \times t$ has to have an equal number per second at each of the two electrodes in a cell, for each electron which is given from an ion to one electrode must be compensated by an ion *receiving* an electron at another electrode. Making the currents at each electrode equal is the role of diffusion. The concentration gradients at the electrodes adjust themselves, in such a way that the rate of arrival of the ions due to the Faradaic transport, together with ions arriving by diffusion and convection, equals the current at each electrode.

2.8. *Non-aqueous solutions*

We have been talking all the time about *water* with ions in it. Ions also exist in other solvents in, for example, alcohol and nitriles. However, there is a tendency for ions in non-aqueous solvents to associate, and therefore the electrical conductivity is often much less than in aqueous solutions.

An exception exists if the phrase ' non-aqueous solvent ' be extended to include molten salts. The situation here is that ionic conductivity is high, and if the molten salt is at a high temperature, 800 K or more, the conductivity is often greater than that of 1 M aqueous solutions.

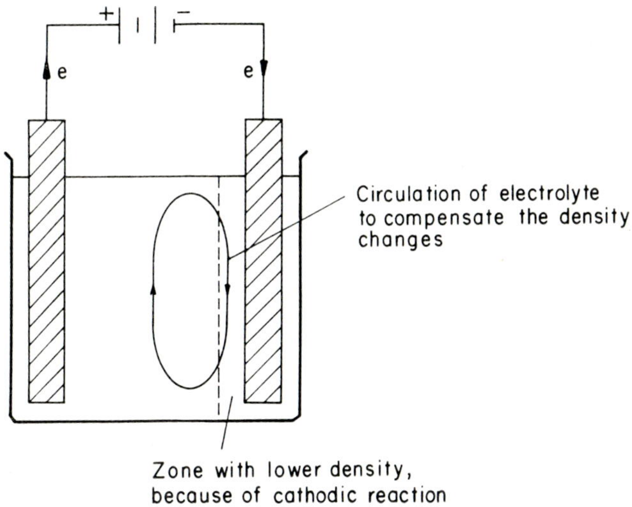

Fig. 2.9. The change in electrolyte concentration produces a flow of electrolyte to compensate the density differences. This type of flow is known as *natural convection*.

2.9. *Protonic mobility*

The proton is not like other ions. It is too small to be treated like an ion of radius 10^{-1} nm or more. In considering the proton, and how *it* conducts, we have to think afresh.

There must be something special about the way in which protons are conducted, for the mobility of the proton (the steady drift velocity in an electric field of $1\ \mathrm{V\ m^{-1}}$) is several times greater than the mobility of any other ion. The mobility of the proton in aqueous solutions is about $36 \times 10^{-8}\ \mathrm{m^2\ V^{-1}\ s^{-1}}$, while that of Na^+ is about 5×10^{-8} and that of Cl^- about $8 \times 10^{-8}\ \mathrm{m^2\ V^{-1}\ s^{-1}}$. In fact, most ions other than H^+ have mobilities of the order of a few times $10^{-8}\ \mathrm{m^2\ V^{-1}\ s^{-1}}$.

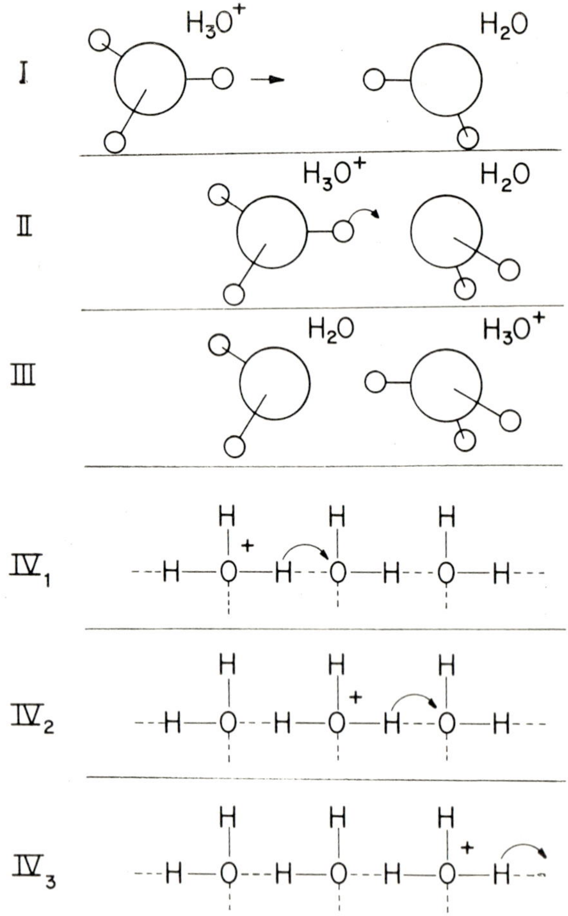

Fig. 2.10. I. H_3O^+ approaches an unfavourably oriented water molecule. II. The coulombic field of H_3O^+ *reorients the water molecule to a favourable configuration.* III. The jump of H^+ converts the H_3O^+ ion into a water molecule, and the water molecule into an H_3O^+ ion. IV. In the presence of an external field the jumps take place preferentially in the direction of the field, and the net result is the migration of H_3O^+.

Why should this be? Other ions 'push' through the solution like little iron spheres being dragged by a magnet through grains of sand—the molecules of the solvent. But the proton doesn't do anything like that. The analogy for *its* movements is a monkey jumping from branch to branch among the trees, swinging on one branch and jumping to another. The proton itself passes easily from one water molecule to the next, so long as the water molecule to which it has to pass is oriented in the right direction. The monkey, hanging from one branch, has to wait until the nearest branch sways back in his direction before he can grab hold of it ; the proton also has to wait, for a bond on an O atom to swing round in the right direction. The process is shown in fig. 2.10.

CHAPTER 3

electric double layers

3.1. *What is an electric double layer?*

WHENEVER two pieces of material (two phases) are brought into contact, a potential difference is set up between them. This occurs through a flow of charge, taking place until the potential difference, $\Delta\phi$, built up by the flow of equalizing charges compensates for the difference of the chemical potentials, $\Delta\mu_i$ (equation 2.12), of individual entities in the two phases. Then, at equilibrium:

$$\Delta\mu_i + z_i F\Delta\phi = 0. \tag{3.1}$$

Equation (3.1) can be put in the form:

$$\Delta\tilde{\mu}_i = \Delta(\mu_i + z_i F\Delta\phi) = 0, \tag{3.1'}$$

where

$$\tilde{\mu}_i = \mu_i + z_i F\phi \tag{3.2}$$

is called the *electrochemical potential* of the species, i. Thus, at equilibrium:

$$(\tilde{\mu}_i)_{\text{phase 1}} = (\tilde{\mu}_i)_{\text{phase 2}}, \tag{3.3}$$

for each species, i.

It is not necessary to devise specific examples of electric double layers, because wherever there is a junction between two pieces of material, involving a layer of moisture or an actual solution, there is an electrochemical double layer. Thus it is virtually impossible for the electrochemical potential of an entity in one phase to be the same as it is in another phase, before the phases are brought into contact; and if the electrochemical potentials are not equal, a current will flow until they become equal, thus developing a potential difference. Nevertheless, a few examples of real double layers are given in fig. 3.1.

Electrochemical science discusses the effects of the transfer of electrons between phases. These effects extend from simple situations like the direct production of electrical energy from hydrogen and oxygen in a fuel cell, to more complex cases, such as the breaking in two of a tanker. Electrical double layers are *everywhere*, and the phenomena of electrochemical science are as widespread as the presence of surfaces†.

† Let us distinguish between inter*faces* and inter*phases*. A double layer has two *faces* which are opposite each other: for example, a silver electrode face and the face of the excess charge which is opposite in solution. Between these faces is the interphase—the zone *between* the two faces.

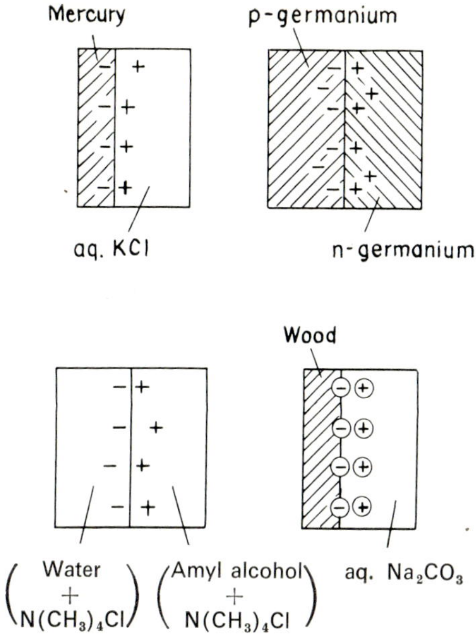

Fig. 3.1. *Above*: A double layer at a metal/ionic solution interface, and a p–n junction. *Below*: Interface potential due to the different solubility of ions in water and amyl alcohol and the double layer at a wood/aq. Na_2CO_3 interface.

3.2. *Polarizable and non-polarizable interfaces*

If we change the number of electric charges which flow per unit time on to a metal surface from an outside source of power, two things can happen. First, these charges can easily leave the surface of a metal in contact with a solution and neutralize protons in the adjacent solution, forming adsorbed atoms of hydrogen on the metal. If there really is no difficulty in this, the only effect of attempting to change the surface charge on the metal (by, say, increasing the flow of electrons from an outside source) will be the formation of more hydrogen atoms at the interface, and nothing else. There would be no change in net charge on the surface, and hence no change in potential between the metal and the solution. The electrons flow in, and the charges on the surface never increase in concentration because the electrons jump across the double layer and form H from the H^+ ions in the solution. The interface is then termed ' non-polarizable '.

On the other hand, there are interfaces which are ' polarizable ' (cf. fig. 3.2). In the ideal *polarizable* interface, when charge flows in from outside it doesn't flow out on to entities in the solution. In such cases there *is* difficulty in the transfer of an electron across an interface

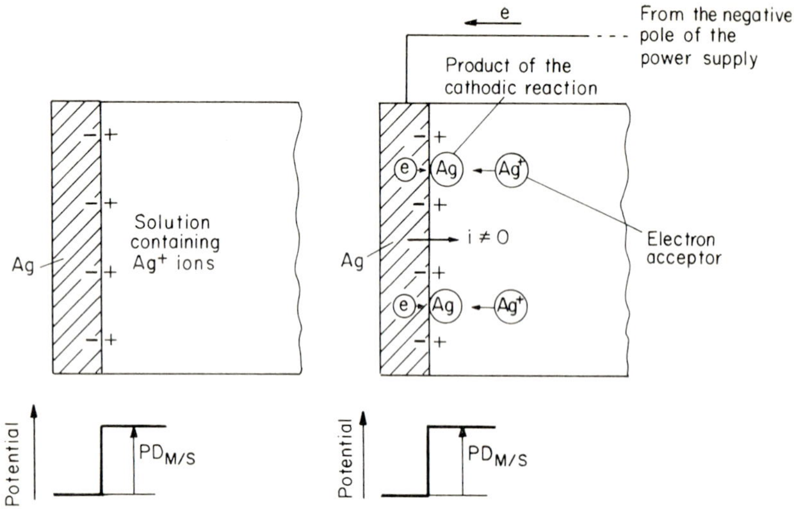

Fig. 3.2. *Left* : a non-polarizable interface with a potential difference $PD_{M/S}$. *Right* : the electrons from the supply flow to the ions of the solution, but the potential difference $PD_{M/S}$ is unchanged.

to H^+ in solution. The electric double layer then acts like a capacitor. One plate is the electrode and the other plate is the layer of charges on the solution. For a *real* capacitor the relation of the potential difference V between its plates to the charge on its plates is given by :

$$V = (d/\epsilon_r \epsilon_0) q \qquad (3.4)$$

where q is the charge per unit area, d is the distance between the plates, ϵ_0 is the permittivity of a vacuum, and ϵ_r the relative permittivity of the material between the plates†.

It follows that for types of surface in solution which are ' polarizable ', if the electric charge is made to increase on one side of the plate and does not leak away into the solution, the potential difference across the double layer will increase or decrease as the charge on the plate increases or decreases. The physical meaning of a polarizable interface is shown in fig. 3.3.

Both types of interface discussed here are idealized extremes, and in reality there are no *completely* polarizable interfaces nor are there any interfaces which are *completely* non-polarizable. However, some interfaces are much more polarizable than others, and to simplify we call some of them ' polarizable ' (though a little charge leaks across them) and some ' non-polarizable ' (though there is a slight build-up of charge and potential across the interface).

† The capacitance per unit area of this parallel-plate capacitor is $\epsilon_r \epsilon_0 / d$.

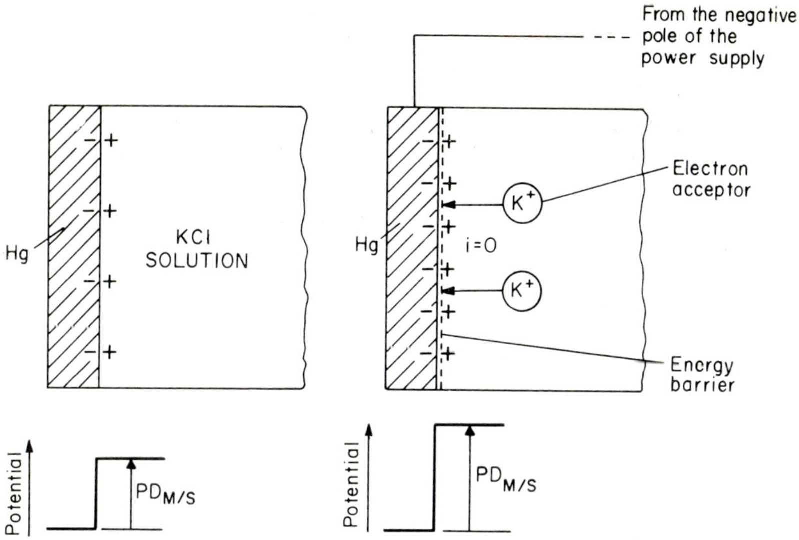

Fig. 3.3. *Left*: a polarizable interface with a potential difference $PD_{M/S}$. *Right*: the electronation reaction does not take place because of an energy barrier. The current is zero, but the potential difference $PD_{M/S}$ grows.

An example of a non-polarizable interface occurs between mercury and a solution containing mercurous nitrate. An example of a polarizable interface occurs between mercury and a solution containing potassium chloride.

3.3. *Measuring meaningful potential differences at interfaces*

(i) The central event (electron ejection or reception by an electron conductor to or from ions in an adjoining solution) in electrochemical science takes place at individual interfaces, and the potential difference between the metal and the solution is of great importance. The difficulty is that we cannot measure the numerical value of it alone, though it will be shown later that we can measure *changes* in it.

We cannot measure the absolute value of a potential because if we try to measure the value of *one* absolute potential difference, we always get involved with another, which is also unknown. This can be seen in fig. 3.4.

The actual measurement which we can make is (from fig. 3.4) always that of the algebraic sum of the potential differences corresponding to these two interfaces; it is never the *value* of a potential difference at just the one interface.

(ii) Can we measure *changes* of potential differences at one electrode–solution interface?

We have to make do with a measurement not of the value of an individual potential difference at the one electrode–solution interface, but with the potential of a *cell* (the sum of at least two potentials). Then, if we keep all the potentials in the cell constant, except one which we vary, the variations of the potential of the *cell* (which we can measure) will be the same as the variations of the single one which we are varying.

Consider fig. 3.5. It shows a working electrode (the central electrode), in a circuit with two other electrodes. One of these is called an *auxiliary* electrode ; it is used to put in and take out charge from the solution, according to the number of electrons flowing from the central power source. Charge thus flows round the circuit between the working electrode (the one we are interested in), the power source, and the auxiliary electrode : the cell is said to ' work '.

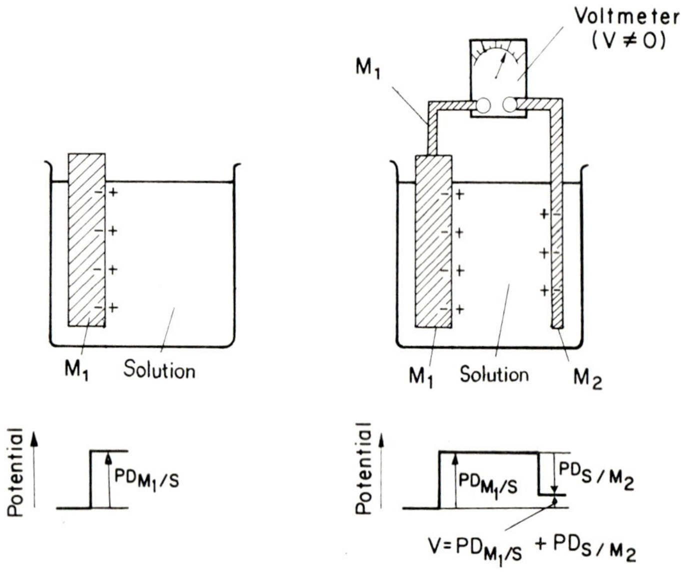

Fig. 3.4. *Left* : the interface between metal and solution, and the potential difference $PD_{M_1/S}$ that we try to measure. *Right* : with a complete cell there is another interface S/M_2 and the voltmeter reads the algebraic sum $V = PD_{M_1/S} + PD_{S/M_2}$.

Now, how can we find the changes in potential difference at the metal–solution interface of the middle electrode, the ' working electrode ' ? We can't measure the *absolute* potential difference between the metal and solution, only its changes. To measure the latter, we introduce a *reference cell.* The sub-cell in fig. 3.5, to the left of the working electrode, consists of a *reference electrode*. A convenient one can be made by bubbling hydrogen over a platinized platinum plate dipping into a molar solution of H^+ ions. The metal–solution interface associated with this reference electrode then has a potential independent of the variations we might make in the potential of the *working* electrode, because it is controlled only by the concentration of hydrogen ions in the solution (we keep that constant) and the pressure of hydrogen above it (kept at a pressure of 1 atmosphere). A very high impedance voltmeter is connected between the reference electrode and the working electrode, so that when we measure the potential difference between

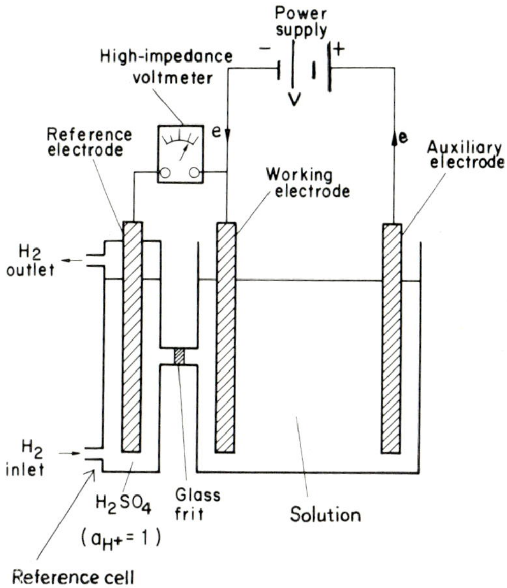

Fig. 3.5. The voltmeter is measuring $V = PD_{\text{ref. elec.}} + PD_{\text{work. elec.}}$. The auxiliary electrode alters the potential difference at the working electrode; $PD_{\text{ref. elec.}}$ is independent of variations in $PD_{\text{work. elec.}}$, and thus $\delta V = \delta PD_{\text{work. elec.}}$.

these two electrodes, exceedingly little current (perhaps $< 10^{-14}$ A) passes. This is necessary because the potential difference between the electrodes is affected if current passes. The potential we want to measure is that of the *working* electrode which corresponds to the current reaching it from the auxiliary electrode, and we don't want any extra current from the reference electrode. By using the very high-impedance voltmeter we make such extra currents associated with the reference electrode so small that their effects are negligible.

What happens when we vary the voltage *V* of the power source in fig. 3.5 ? As more electrons pass in from the auxiliary electrode, they accumulate on the working electrode and its potential changes (equation 3.4).

We note the change in reading of the voltmeter, measuring the potential of *the cell* (working electrode—voltmeter—reference electrode). Since the potential differences associated with the reference electrode are unaffected by whatever is happening to the working electrode, changes registered by the voltmeter will represent the changes in the potential of the working electrode as more electrons reach it.

(Potential measurements in electrochemical science are always measurements of *changes* in *potential* or of *differences of potential.*)

There are two very special potentials (rather, differences in potential) which are measured in this way.

One of these is called the '*thermodynamic reversible potential*'. It is the potential of an electrode measured when the ions in the solution and the electrons in the metal are in equilibrium. Electron transfer is still occurring at an interface in equilibrium, but if the metal concerned is silver and the ions in solution Ag^+, then at equilibrium electrons transfer to Ag^+ ions in solution, forming Ag, at the same rate as that at which Ag^+ ions are being formed from Ag. $Ag^+ + e \rightarrow Ag$ has the same velocity as $Ag \rightarrow Ag^+ + e$.

But we must not pass significant currents through such electrodes, because the electrode will then depart from the non-polarizable condition: 'overpotential' will arise (Chapter 4), so that the potential will no longer correspond to that at 'thermodynamic equilibrium'.

How is a numerical value for the thermodynamic reversible (equilibrium) electrode potential obtained ? We have to measure a *cell* potential again. We can get values of reversible potentials of different metals which are comparable if we always make *one* electrode in the cell the *same* one. Then, as we change the other electrode in the cell, while keeping the standard reference electrode the *same*, we consider the difference of the cell potentials as the difference between the two individual 'reversible' electrodes when they are in contact with molar solutions containing their ions (fig. 3.6).

Another special kind of potential, the *potential of zero charge*, is the potential exhibited (and measured in a cell such as that shown in fig. 3.5) when an electrode is equally balanced between its tendencies to have an excess negative and an excess positive charge, and therefore does not have a charge on it. The significance of this potential was made clear by Frumkin. We shall talk about it a little later in this chapter.

3.4. *The electric field strength in the electrode double layer*

A key to understanding the part played by electrochemistry in so many phenomena is to appreciate the considerable electric field strength in the interphase double layer. By taking an approximate value of the potential difference within the double layer to be 1 V, and an approximate value of the distance between the layers of charges in the double layer to be about 0·3 nm, we find the electric field strength to be

$$1/(0{\cdot}3 \times 10^{-9}) \simeq 3 \times 10^{9}\ \mathrm{V\ m^{-1}}.$$

An electric field strength as high as this is not obtainable except in electrochemical interphases. It is this strong field acting upon molecules and ions during their electrochemical reactions which so greatly affects their reaction rates, the *rates* of processes at electrodes.

3.5. *What is adsorption?*

We know what adsorption in classical chemistry is. Hydrogen gas will adsorb on tungsten, forming a layer of hydrogen atoms closely bound to tungsten.

In such cases, adsorption happens suddenly. The hydrogen gas,

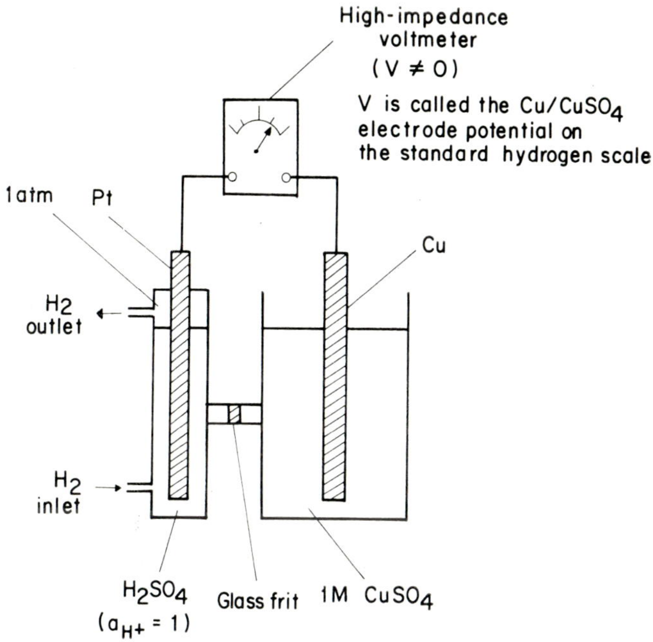

Fig. 3.6. Absolute electrode potentials cannot be measured, but they can be *compared* by coupling them with the same reference electrode (e.g., a standard hydrogen electrode) and by comparing the potentials of the complete cells thus obtained.

when it is more than about 0·5 to 1·0 nm from the tungsten, is unaffected by the presence of the tungsten ; it is only when the hydrogen molecules actually strike the tungsten that they dissociate and H atoms bind with the tungsten atoms.

The situation is different in electrochemical adsorption. In electrochemical systems the coulombic electrical forces which stretch out from the metal to affect the solution are much longer in range than the chemical forces which affect molecules in chemisorption.

Thus, if we take the point of view of an ion swimming towards the interface, it would observe a significant change in concentration of ions in the solution before the metal is reached. Adsorption, in the electrochemical situation, is in the *interphase*, not simply on the surface. Hence, adsorption has a *more general* meaning in solution than it does for a gas–metal interface, where it merely refers to a layer of gas molecules adsorbed on the metal. The concept of an interphase has little meaning for gases because the gas–metal change in phase is sharp ; the interphase has a negligible thickness.

The region affected by electrical forces at interphases, namely the breadth of the interphase, is dependent on the concentration of the ions in the solution with which the electrode is in contact. It will be at least 0·5 nm but becomes much more, as much as 100 nm, when the concentration is very low. Consider a solution of uni-univalent electrolyte. The thickness of the region affected by electrical forces from the electrode is given by the same formula (2.15) which gave the thickness of the ionic cloud around a central ion :

$$\kappa^{-1} = 1/B\sqrt{c} \tag{3.5}$$

the value of B being dependent on the temperature and the electrolyte permittivity. For $c = 10^{-4}$ mol dm^{-3} and $T = 298$ K, this formula gives $\kappa^{-1} = 30$ nm.

3.6. *There are two kinds of adsorption in electrochemical situations*

In systems connected with adsorption of gases on metals, there are two kinds of adsorption. In one kind the forces which take part are non-chemical ones, for example dispersion forces.† A weak physical adsorption takes place, as for example with argon on potassium.

In the other type of adsorption (earlier exemplified by hydrogen on tungsten) molecules break into two parts and gas atoms are adsorbed individually, with bonding to the metal atoms. This is called chemi-

† There exists an attractive force, involving energies of the order of kT (k is the Boltzmann constant), between ALL molecules, even between the molecules of inert gases like He. These dispersion forces (van der Waals forces) are very weak, though they become significant for very small separation distances, especially at low temperatures.

sorption, because a chemical bond forms between the atom and the metal.

The electrochemical situation also shows two kinds of adsorption. The kind of adsorption similar to physical adsorption is called coulombic. If there is an excess charge of say $+0{\cdot}1\ \mathrm{C\ m^{-2}}$ on the metal, then there will be an excess charge of $-0{\cdot}1\ \mathrm{C\ m^{-2}}$ on the solution, which means an excess of anions in the solution. This solution charge will be made up of some anions and some cations. Thus, there will be somewhat less than $0{\cdot}1\ \mathrm{C\ m^{-2}}$ of actual *negative* charge on the solution side, because there will be some *repulsion* of positive charge away from the electrode, which is equivalent to the presence of some excess negative charge. The situation is shown in fig. 3.7.

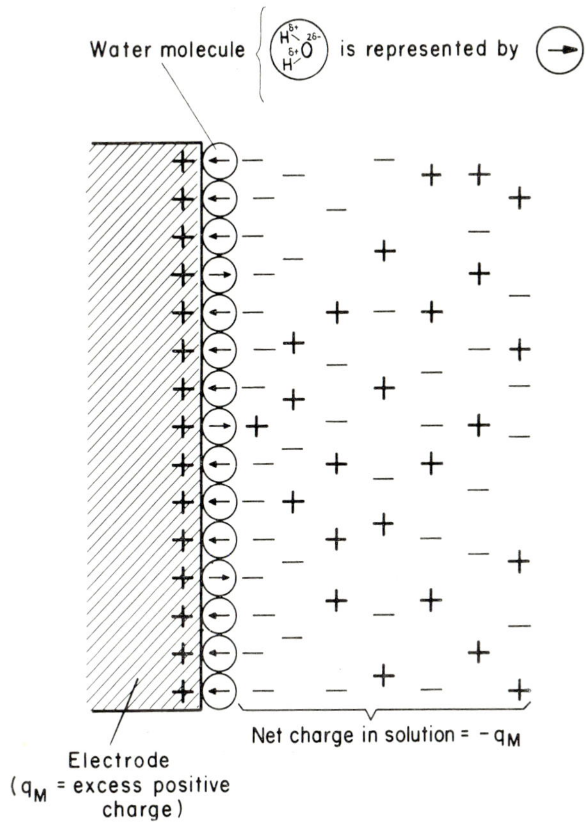

Fig. 3.7. The physical (coulombic) adsorption at a metal/NaF aq. solution interphase. The fluoride anions are attracted and the sodium cations are repelled. The *excess* charge of the metal and the *net* charge on the solution side are equal in magnitude, but of opposite sign. A layer of H_2O molecules resides on the electrode surface.

A characteristic of coulombic adsorption is that no ions are in direct physical contact with the metal. The excesses and deficiencies of ions occur in the solution, in the interphase. A layer of water molecules remains on the surface of the electrode and divides the electrode–metal from the solution (fig. 3.7).

In the other kind of adsorption ions do actually come into contact with the electrode (fig. 3.8). *Contact* adsorption is analogous to chemisorption. There is not necessarily a chemical bond between the contact-

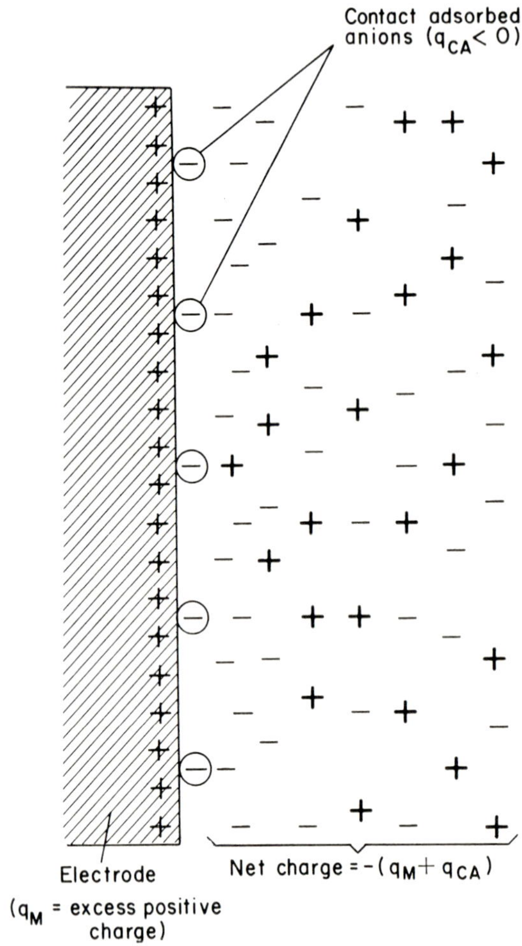

Fig. 3.8. Interphase with both contact and coulombic adsorptions. The excess charge of the electrode is equal and of opposite sign to the algebraic sum of the contact-adsorbed anions charge and the net charge in solution. (Adsorbed water molecules not shown.)

adsorbed species and the metal. We are not sure what kind of a bond it is. The point is that ions taking part in it contact the electrode and remain attached. Of course, the rules remain that the excess net charge on the adsorption side must be equal to the excess net opposite charge on the metal. However, there are now three components of the adsorbed charge (see fig. 3.8):

Contact adsorption (i.e. the anions in contact with the electrode); anions still in the diffuse layer; and cations which are also in the diffuse layer. The net charge on the diffuse layer is negative if $q_M + q_{CA} > 0$ (as in fig. 3.8), and positive if $q_M + q_{CA} < 0$.

We can measure these contributions and distinguish between them all. The way we do it is complicated and we shall not work it out here. By making measurements of the surface tension of the metal–solution interface as a function of the total charge on the metal at various concentrations, it is possible to use thermodynamic and other equations to give values to the two types of adsorption which we have described, i.e. physical and contact adsorption. Radio-tracer experiments, and some optical methods, can increase our knowledge of adsorption, and these types of measurement are more general because they do not need surface tension measurements, which are only easy for the liquid metals.

For several metals, and a variety of solutions in contact with them, the magnitude of the adsorption of ions as a function of electric charge (at one concentration) and as a function of concentration (at one charge) is known.

3.7. *The potential of zero charge*

We can talk about a metal having either an excess positive or an excess negative charge. We can take electrons out of it, so giving it

Metal	Electrolyte	Concentration /(mol dm^{-3})	pzc/V†
Aluminium	KCl	0·01	−0·52
Bismuth	KCl	0·01	−0·36
Cadmium	KCl	0·01	−0·92
Lead	KCl	0·01	−0·69
Cobalt	Na_2SO_4	0·01	−0·32
Copper	Na_2SO_4	0·01	+0·03
Gold	Na_2SO_4	0·01	+0·23
Silver	Na_2SO_4	0·05	−0·07
Mercury	NaF	0·10	−0·19
Iron	H_2SO_4	0·005	−0·37

† The values are with respect to the standard hydrogen electrode.

Table 3.1. The potential of zero charge for several metals.

excess positive charge (without further electrons leaking across the double layer). This will cause a rearrangement of charge on the solution side, and this rearrangement may involve partly the contact-adsorbed material and partly the coulombically adsorbed material referred to above.

In an experimental situation such that there is a negligible contact adsorption (and that is so for all systems at sufficiently low electrolyte concentration), the potential at which there is zero charge on a metal, ' the potential of zero charge (pzc) ', is characteristic of the metal alone.

Such a potential is obviously important and values for various metal/electrolyte systems are given in Table 3.1.

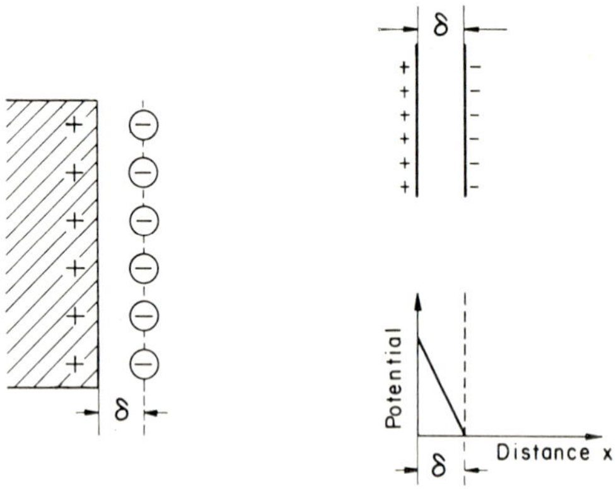

Fig. 3.9. The Helmholtz model. *Left* : the double layer consists of two sheets of charge, one on the electrode and the other in the solution at a distance δ. *Right* : the double layer is equivalent to a parallel-plate capacitor ; the potential varies linearly with the distance x : $\psi = \psi_0(1-(x/\delta))$, where ψ_0 = the potential at the electrode surface.

3.8. *Models for the double layer*

It will help us to understand things if we proceed by stages towards a model of the double layer which is nowadays considered to agree with the experimental observations. Historically, the first model was that of Helmholtz and Perrin (fig. 3.9). It regards the double layer simply as a plane layer of charges (excess adsorbed ions) in the solution, confronting a plane layer of charges of opposite sign on the electrode.

It was from this simple idea that the name ' double layer ' developed.

The Helmholtz model is much oversimplified. A strict and rigid distribution of ions on a simple plane seems unlikely. There must be thermal agitation of the ions on the solution side and they are not confined to one single sheet at 0·5 nm or so from the metal, but distributed over a much greater distance into the bulk of the solution.

This idea of the distribution of charge on the solution side was the main point of another theory, the Gouy–Chapman theory of the diffuse layer. It resembles, in respect of this distribution, the Debye–Hückel theory of the ionic atmosphere (Chapter 2), which it preceded historically. The interplay of thermal and electrical forces near a charged electrode builds up an ionic cloud of effective thickness κ^{-1}, where κ^{-1}—in the case of a 1,1-valent electrolyte—is given by (3.5), which becomes, when we write B out explicitly :

$$\kappa^{-1} = 1/B\sqrt{c} = 1/[(N_A e_0^2/500\, \epsilon_r \epsilon_0 kT)^{1/2}\sqrt{c}]. \qquad (3.6)$$

In this relation, c is the bulk concentration of the 1,1-valent electrolyte in mol/dm^{-3}, and k the Boltzmann constant ($1{\cdot}38 \times 10^{-23}$ J K^{-1}).

Let us consider the case of a 1,1-valent electrolyte, and suppose that the electric potential, ψ_0, at the electrode surface is small.

Let us also simplify the model by assuming that there are no *contact*-adsorbed ions on the electrode. In that case, the excess charge on the electrode is equal to the total charge held in the diffuse layer on the solution side. The Gouy–Chapman theory gives the following expressions, under these conditions, for the variation of potential with distance x from the electrode, and for the charge density q_d on the solution side (equal in magnitude, but opposite in sign to the charge density, q_M, on the electrode surface) :

$$\psi_x = \psi_0 \exp(-\kappa x) \qquad (3.7)$$

$$q_M = -q_d = (\epsilon_r \epsilon_0/\kappa^{-1})\psi_0. \qquad (3.8)$$

If we compare equations (3.8) and (3.4), we see that the double layer can be simulated by a parallel-plate capacitor with the plates separated by the distance κ^{-1}. This is an important conclusion about the meaning of κ^{-1}.

The concentration profiles within the diffuse layer are given by the following exponential relationships :

$$c_+ = c \exp[-(e_0\psi_x/kT)]$$

$$c_- = c \exp[+(e_0\psi_x/kT)] \qquad (3.9)$$

where c_+ and c_- are the concentrations of cations and anions at a distance x from the electrode, respectively, and ψ_x the potential at distance x from the electrode. Figure 3.10 shows the concentration profiles within the diffuse layer.

The principal conclusions of the Gouy–Chapman theory are summarized in fig. 3.11.

The Gouy–Chapman theory gives an overestimate, with too much stress on *distribution*, which means physical (coulombic) adsorption,

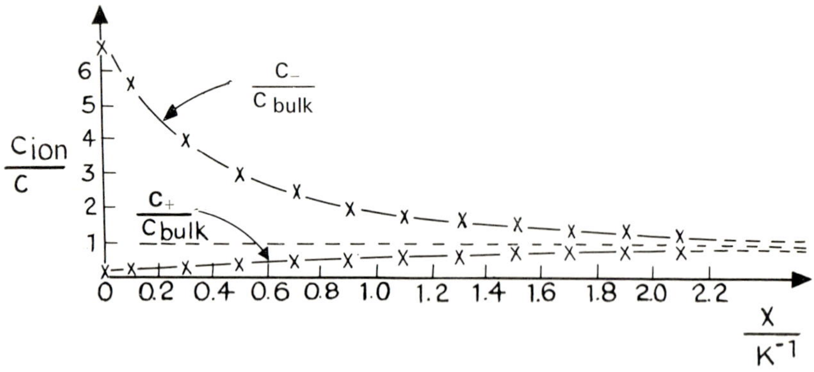

Fig. 3.10. The concentration profiles within the diffuse layer at a metal/NaF interphase. Distance is expressed in κ^{-1} units. The calculation has been carried out by letting $\psi_0 = 50$ mV, and $T = 298$ K. c_- stands for the concentration of F^- ions in the solution if the electrode is positively charged. c_+ then represents the Na^+ concentration. If the electrode is a cathode, the F^- anions are repelled while the Na^+ cations are attracted ; c_- then indicates the Na^+ concentration while c_+ stands for the concentration of F^-. All concentrations are expressed as fractions of the concentration c_{bulk} in the bulk of the liquid.

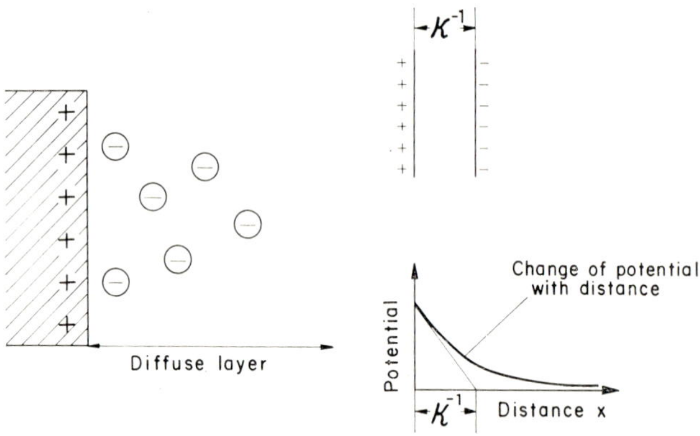

Fig. 3.11. The Gouy–Chapman model. *Left* : the net diffuse charge density is scattered in the solution under the influence of thermal and electrical forces. *Right* : the double layer equivalent to a charged parallel-plate capacitor, distance between the plates equal to κ^{-1} ; the potential varies exponentially with the distance ; $\psi = \psi_0 \exp(-\kappa x)$; in the analogy with a parallel-plate capacitor, this is replaced by the linear approximation : $\psi = \psi_0(1 - (x/\kappa^{-1}))$.

and not enough stress upon the sticking of the ions on to the electrode on the solution side.

A third and better model of the situation in the double layer is that of Stern, in which the ideas of Helmholtz and of Gouy–Chapman were combined. According to Stern, some of the charge on the solution lies in a plane called ' the plane of closest approach ', and the remainder is spread out in the solution (in Gouy–Chapman style). The plane of closest approach of the ions to the electrode divides the solution side into two regions : a Helmholtz–Perrin region, and a Gouy–Chapman region. Each of these two regions can be simulated by a parallel-plate capacitor (see fig. 3.12), and therefore the double layer is equivalent

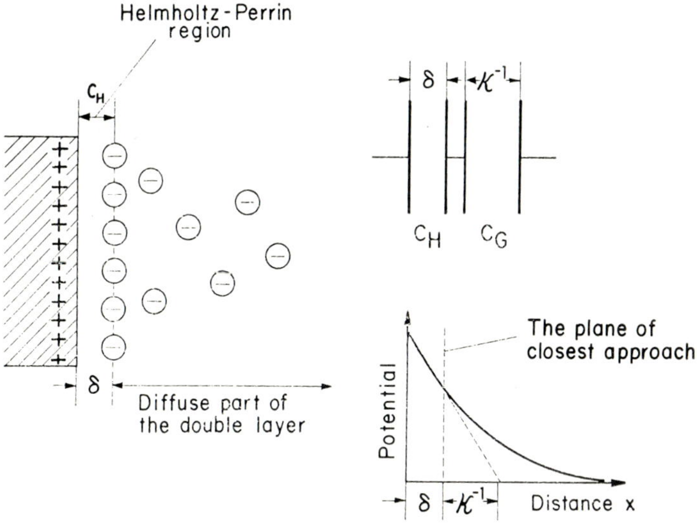

Fig. 3.12. The Stern model. *Left* : the double layer can be divided into two regions : δ is the closest distance of approach of the ions to the electrode. *Right* : the double layer is equivalent to two capacitors in series ; C_H that of the Helmholtz–Perrin region, and C_G the capacitance equivalent to the diffuse layer.

to two capacitors in series. The variation of potential with distance is linear in the first region and exponential in the second region (fig. 3.12).

The total capacitance of the double layer is therefore given by :

$$1/C = 1/C_H + 1/C_G \tag{3.10}$$

where C_H and C_G are the capacitances of the two equivalent capacitors.

Let $\psi_{x=0}$ be the potential of the electrode with respect to that of the bulk of the solution ; and consider the case when ψ_0 (the potential of the Helmholtz plane) is sufficiently small. Then (equation 3.7) the potential variation can be described by :

$$\psi = \psi_{(x=0)} \frac{x}{\delta} \quad \text{for} \quad x < \delta$$

$$\psi = \psi_0 \exp\left[-\kappa(x-\delta)\right] \quad \text{for} \quad x > \delta. \tag{3.11}$$

The fact that water molecules in contact with the electrode will no longer have the same structure as water in the bulk of the solution, but will be oriented as dipoles on to the electrode (thus contributing to the potential difference across it), was neglected until the 1960's. A scheme which takes this important factor into account is shown in fig. 3.13.

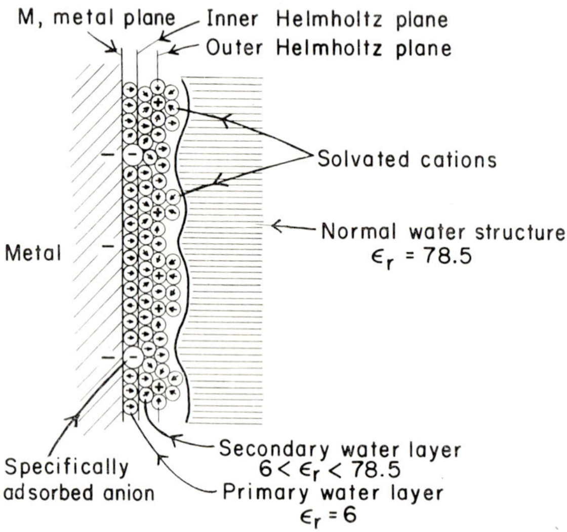

Fig. 3.13. The structure of the double layer according to Bockris, Devanathan and Muller.

As we can see there, a distinction is made between contact-adsorbed ions and the solvated cations which are adsorbed out of direct contact with the electrode. The plane passing through the centres of contact-adsorbed anions is called the ' inner Helmholtz plane ' and that passing through the centres of solvated ions, the ' outer Helmholtz plane '. In fact, there are two outer Helmholtz planes, one for the cations and the other for the anions, but in our first-approximation discussion it was assumed that these two planes were identical. Thus, three regions must be distinguished in the structure of the solution side at an electrified interface. The water layer in the first region is stressed by the force exerted on it by the electrode field, and thus the relative permittivity of this ' primary water layer ' with fully oriented water molecules is much less than the permittivity of water in bulk, and

ϵ_r = about 6. (This is the relative permittivity of water molecules when they are dielectrically saturated, as they will be at the high field strength attained in double layers.)

The water molecules in layers further from the electrode than the first are oriented by the electrode field, but are also partly disoriented by thermal and hydrogen-bonding influences. The relative permittivity of this secondary water layer rises with increasing distance from the electrode, from about 6 to the bulk value of 78·5 at 298 K.

This is the model of the double layer that we shall use (fig. 3.13). But when making rough and ready calculations we may consider the model in a simple Helmholtzian way (fig. 3.9).

3.9. *Adsorption of organic molecules at electrodes*

So far, we have discussed two types of ion adsorption. Some ions are adsorbed in actual contact with the electrode. And some ions are content to stay out in the solution, although the fact that they don't have the same concentration there as they would if the electrode field were not acting on them makes us call them adsorbed ions too.

Organic molecules, for example alcohols, are also adsorbed upon electrodes: an example of this is shown in fig. 3.14. We can understand this in a rough and ready way by making three simplifying assumptions. Suppose the organic molecule does not dissociate upon adsorption. This is true for e.g., aliphatic alcohols on mercury. Suppose also that there is nothing else involved except electrostatic forces (this is a big assumption and will be quite wrong for, say, the adsorption of aromatic molecules on mercury). Finally, suppose that these forces affect the water molecules considerably but the organic molecules only to a negligible extent. This will be true if the organic molecule is oriented to the electrode so that its dipoles are in the Gouy region, where the field from the electrode is small enough to make the energy of interaction of it with the organic molecule negligible.

When these grossly simplifying assumptions apply, it will be largely the electrical behaviour of the water molecules which will determine the adsorption of the organic molecules. When the excess charge on the electrode is highly positive or highly negative, the water molecules will be more strongly adsorbed than when the charge is near zero (because the water molecule is assumed to have a much stronger dipole moment than that of the organic molecule). Of course, the water dipoles will have to turn around when the sign of the charge on the electrode changes.

From these considerations the expected *trend* is clear. It is near the potential of the zero charge (pzc) for the metal concerned that one expects maximum adsorption of the *organic molecules*; and the adsorption–charge density relation will tend to be parabolic.

The competition between water and organic molecules is shown in fig. 3.14.

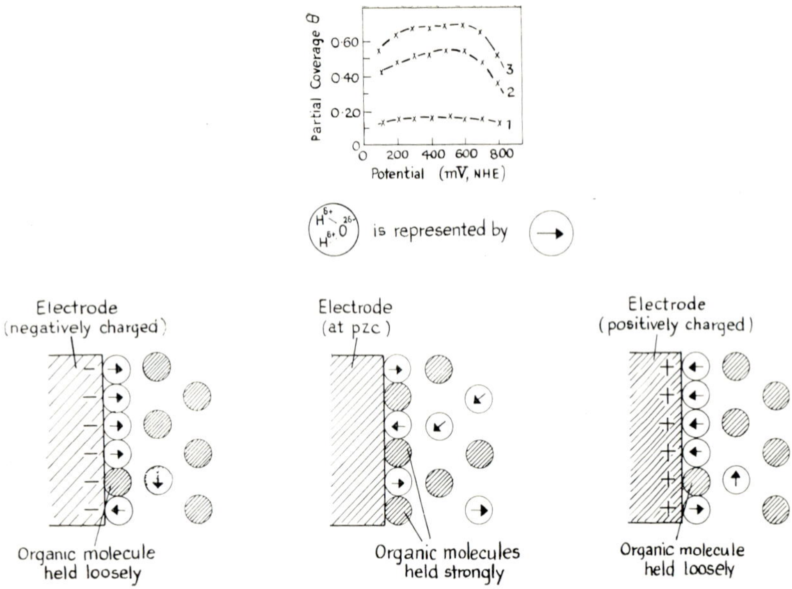

Fig. 3.14. The adsorption of benzene on a Pt electrode from 1 M H_2SO_4, at 323K, for three concentrations: (1) $2{\cdot}8\times10^{-7}$ mol dm^{-3}; (2) 11×10^{-7} mol dm^{-3}; and (3) 70×10^{-7} mol dm^{-3}. (After: W. Heiland, E. Gileadi and J. O'M. Bockris, *J. Phys. Chem.*, **70**, 1208 (1966)).

All this is only true in simple situations, and there are often other interactions which become important. Thus, Russian authors, particularly Frumkin and Damaskin, have stressed the strong electrical interaction between the *organic* molecules on the surface, and the change of the shape of the interface connected with it. They have neglected, however, explicitly to consider the independent effect of water molecules on the situation. Their view would be correct if the electrical effect of the organic–electrode interaction were much greater than that of the water molecules, as might perhaps sometimes be so.

A complication arises when the organic molecule dissociates in the same step in which it is being adsorbed. This is the case when propane is adsorbed on metal in acid solution. When propane is added to phosphoric acid in which a platinum electrode is immersed, a current is generated showing that when the propane is adsorbing there is a *dissociation* reaction which yields charges, probably according to:

$$C_3H_8 \rightarrow (C_3H_7)_{ads} + H^+ + e.$$

In such cases, the treatment of adsorption is more complicated and the competition between water dipoles and organic molecules will become a smaller part of the story.

CHAPTER 4

the kinetics of electronic charge transfer

4.1. *Introduction*

In the last chapter, we described the interphase, the region between the two sets of charge which make up the electrical double layer. We also described how the electric field in the interphase varies with distance from the electrode.

In this chapter, we describe the act of charge transfer and relate the rate of that transfer to the potential difference in the interface.

4.2. *The physico-chemical picture of charge transfer*

The formula for the probability, P_E, that an energy level E in a metal will be populated by electrons is known as *the Fermi–Dirac distribution function*:

$$P_E = \frac{1}{\exp[(E-E_F)/kT]+1} \tag{4.1}$$

where E_F is known as the *Fermi energy* and k is the Boltzmann constant.

Let us set $E=E_F$ in equation 4.1. Then the exponential term becomes $e^0=1$ and $P_E=\frac{1}{2}$, so the Fermi level is that energy level which is half-populated by electrons at $T>0$ K.

When $P_E=1$, the energy level of energy E is full, and the electrons find no states available to move into in the metal. *Levels which are full will not conduct electricity* because they have no empty sites into which moving electrons can pass. Thus, electron levels which are full cannot act as source or sink for electrons and cannot take part in an electrode reaction.

Correspondingly, when P_E tends to zero, there are not enough electrons present to act as a source of electrons in charge transfer.

Let us consider an electron emitted from a metal into a vacuum. As the electron tries to enter the vacuum a positive *image* charge is induced on the metal, and the electron experiences a force pulling it back. Thus, the electron encounters an energy barrier of finite height and width.

We should expect, using classical mechanics, that the electron would have to have an energy greater than the energy of the barrier to leap over it and escape from the metal. Only a very small fraction of the electrons inside the metal has an energy greater than or equal to that corresponding to the top of the barrier. Consequently, the emission current by such a classical path is very small. By raising the temperature, the exponential exp $(-\Delta E/kT)$ (where ΔE is the barrier height) increases, and an increasing number of electrons pass *over* the barrier. This is what happens when a hot wire emits electrons, as in a classical diode.

The theory of this kind of emission current was given long ago. It is easy to show that, for electrochemical situations, current densities arising in this way at electrodes would be $< 10^{-16}$ A m^{-2}, whereas the normal order of electrode currents is about 10^{1} A m^{-2}.

Thus, the classical *over the barrier* calculation gives results quite remote from currents observed experimentally, and one must conclude that classical mechanics cannot apply to the transfer of electrons from a metal in solution to neighbouring ions. This means that common sense (based on understanding of the way *big* objects behave) does not apply to the electron-level events which occur at electrodes. What *does* apply is quantum mechanics, the physics of the not-so-common-sense behaviour of very small particles. The laws of quantum-mechanical ' tunnelling ' will govern when an electron can leave a metal and go to the solution. And quantum-mechanical tunnelling will determine the probability that an electron can leave the solution and enter the metal.

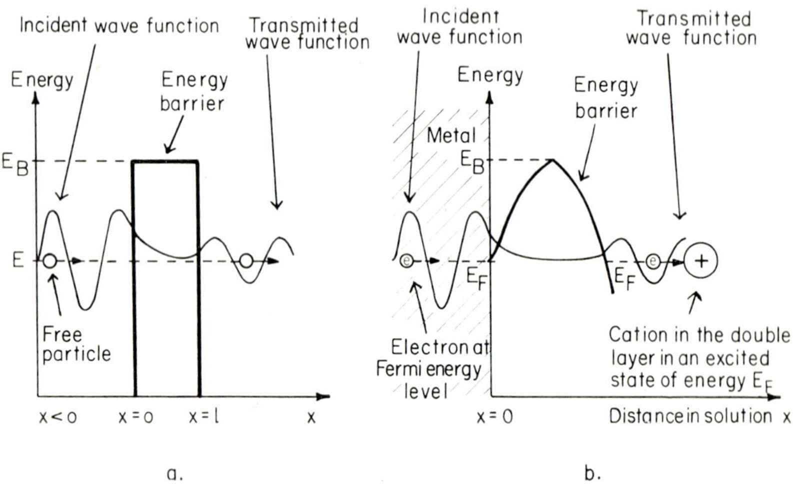

Fig. 4.1. Pictures the quantum-mechanical tunnelling (*a*) of a free particle through a rectangular energy barrier in a vacuum, and (*b*) of an electron through a metal–electrolyte interface. In the latter case, tunnelling can be towards only those particles in the solution which are in excited states, and which possess an electronic energy level which has the same energy, E_F, as the electrons in the metal.

Figure 4.1 (*a*) illustrates quantum-mechanical tunnel transfer through a barrier. According to classical mechanics, a free particle of energy E smaller than the energy E_B of the barrier could never escape to the region $x > l$. In quantum mechanics, there is some probability that the particle will get to the other side of the barrier, because the behaviour of the particle is described by a wave-function.

In fig. 4.1 (*a*) the wave function of the system is superimposed on the energy diagram and the whole picture is intended to be interpreted qualitatively. The behaviour of a particle can be predicted in quantum mechanics if the behaviour of the wave function is known. In the first region, $x < 0$, and beyond the barrier the wave function varies periodically with distance, x. In the second region ($0 < x < l$) the wave function varies in a non-periodic way. There is a formula for the ratio between the number of particles penetrating the barrier at $x = l$, and the number of particles striking the barrier at $x = 0$. The essential part of this formula is the exponential term :

$$\exp\{-4\pi l h^{-1}[2m(E_B - E)^{1/2}]\} \qquad (4.2)$$

where m is the mass of the particle. Hence, the transmission coefficient will not be zero unless at least one of the quantities E_B, l, or m has the value infinity. For a given $(E_B - E)$ and l, the transmission coefficient *increases* as the mass of the particle *decreases*. Thus, electrons tunnel more easily than atoms. But it is electrons which have to cross the barrier.

It is easy to show, from equation 4.1, that if E is about 0·1 eV *above* the Fermi energy the probability (at 298 K) of the level containing an electron is $< 10^{-2}$, and if it is 0·1 eV *below* the Fermi level, the probability that the level will be full is 0·99. We shall therefore call a level above the Fermi level ‘ empty ’ and a level below the Fermi level ‘ full ’.

Thus, for energy levels around the Fermi level, there are many electrons available to react with materials in solution. For levels below the Fermi level, there are no empty states and no mobility, so that electron transfer is not possible. We therefore assume (as a reasonable approximation) that all the electrons involved in the electric current do have energies of E_F.

Figure 4.1 (*b*) illustrates tunnelling at a metal–electrolyte interface. The shape of the barrier varies with distance in a *smooth* fashion. Nevertheless, the barrier is qualitatively similar to the rectangular barrier shown in fig. 4.1 (*a*), and for the purpose of our present argument we can safely work in terms of the simpler square barrier.

The tunnelling of an electron (of energy E_F) from a metal to a particle (e.g. a cation) in the solution takes place only if another energy condition is fulfilled. There must be a level in that particle which is available (empty) and which is *at the same energy*, E_F, as the electron in the metal. No energy change is observed during electron transfer, so that electrons must pass from metal to solution, and vice versa, without loss of energy. Hence, we are interested in the probability of excitation of energy levels in the solution particles (e.g. H_3O^+) to the level of the metal's Fermi level. Between *these* levels, and the corresponding levels in the metal, transfer will occur. In the electronation reaction, $H_3O^+ + e \rightarrow H_{ads} + H_2O$, the H–O bonds are excited, and may be *longer* than bonds in the *un*excited state.

4.3. *Rate of transfer of atoms per unit area per second is equivalent to current density in* A m^{-2}

In chemistry, we express rates of interfacial reactions in mol m^{-2} s^{-1}.

In electrochemistry, we express rates in A m^{-2}.

The two expressions are, of course, related. According to Faraday's laws of electrolysis, when we pass 96 500 C ($=F$) across an interface, one mole of material is liberated if the charge on the ionic species involved is e, and $1/z$ moles if the charge is ze.

For current density i in A m^{-2} ($=$C s^{-1} m^{-2}) the number of moles discharged per (metre)^2 per second is i/zF. So the reaction rate is

$$v = i/zF. \tag{4.3}$$

Since i is proportional to v, whenever we mention current-density in electrochemistry we are really thinking of the rate of an electrode reaction. A high current density means a high electrochemical reaction rate.

A typical reaction rate is 10 mA cm^{-2} = 100 A m^{-2}; when the actual calculated value of v will be $100/96\,500 \simeq 10^{-3}$ mol m^{-2} s^{-1} for ions of unit charge.

4.4. *Electrochemical reaction rates are affected by the potential gradient in the double layer*

In electrochemical science, we can affect the *rates* of electron transfer across interfaces—send them in one direction or the other, or make them fast or slow—by changing the electric field in the double layer. It follows that we can affect the rates of electrochemical reactions, for example the rate of electrochemical oxidation of ethylene to CO_2: $C_2H_4 + 4H_2O \rightarrow 12H^+ + 2CO_2 + 12e$.

Before we see why this is so, let us reflect on what a useful thing it is. We can change the speed or the direction of a reaction by means of a switch in the circuit.

What is the electric field strength in the double layer? We have worked it out in one case (page 41) and shown that its magnitude is high, about 10^9 V m^{-1}. Since the distance over which the potential drop at an electrode–solution interface occurs (the distances from the metal to the nearest layer of ions) is not going to change significantly when there is a change of potential on the electrode, the change of potential difference is what mainly determines the field strength: *field strength* $= -$(*potential difference across the double layer*)/*distance.* The potential of the solution is constant, independent of that of the electrode, and so the field strength in the double layer is proportional to the electric potential of the electrode. Roughly, then, when one changes the electrode potential, the change in electric field strength in the double layer is proportional to that change in potential.

4.5. *Why does a change in potential in the double layer cause a change in electrochemical reaction rate?*

When we make an electrode more negative, the Fermi level of electrons in it is raised, and the tendency of electrons to transfer into electronic states possessed by particles in solution is increased. Qualitatively, therefore, the more negative an electrode the faster will be the rate of escape of electrons from the metal to the solution, and the faster the electrochemical reaction.

Likewise, if we make the electrode's potential more positive, there will be more electronic states of ions in solution which overlap with the energy of electrons in unfilled states in the electrode, and the de-electronation reaction is speeded up.

There are two terms associated with these current densities: (1) the cathodic current density, i_-, is the current corresponding to electrons being ejected from the electrode to the solution; (2) the anodic current density, i_+, for electrons ejected by the solution to the electrode.

Thus, for electrochemical reactions driven by outside power sources, the more negative the potential of an electrode the greater the *cathodic* current density; and the more positive the potential of an electrode, the greater the *anodic* current density.

For a better understanding of this conclusion, consider the special case of electrons tunnelling from a metal to hydrated protons in the cathodic (electronating) reaction:

$$M(e) + H^+{-}OH_2 \rightarrow M{-}H + H_2O.\dagger \qquad (4.4)$$

We have seen that radiationless tunnelling can occur only if the electrons can transfer from the Fermi level of the metal to hydrated protons which possess electronic states of the same energy, E_F, as the electrons in the metal. Correspondingly, the radiationless tunnelling condition is fulfilled when the energy of the system before the electronation reaction (i.e. $M(e) + H^+{-}OH_2$) is equal to the energy of the electronated form of the system (i.e. $M{-}H + H_2O$).

What terms are implied in the energy of the system $M(e) + H^+{-}OH_2$? Firstly, the energy of an electron in the metal (i.e. E_F), and then the energy of a hydrated proton. The latter, using the energy of a hydrogen

† The hydrated proton $H^+{-}OH_2$ is also written H_3O^+ and H^+_{aq}.

atom as reference, can be divided into two parts : the ionization energy, to get an H^+ ion, and the heat of hydration, $\Delta H_{hyd.\,H^+}$, to get a hydrated proton $H^+–OH_2$.

The energy of the system $M–H+H_2O$—also with reference to the hydrogen atom—has two terms : the adsorption energy, A, i.e. the energy released when a neutral H atom establishes a bond with the metal, and the repulsion energy, R, between the H atom and the H_2O molecule.

Now, consider that the H_2O molecule of the hydrated proton remains at a fixed distance from the metal, and that the proton oscillates between the metal and this H_2O molecule. Then $\Delta H_{hyd.\,H^+}$ and, hence, also the energy of the system $M(e)+H^+–OH_2$, changes periodically.

Similarly, the movement of the H atom between the metal and the water molecule will change both A and R terms, and hence the energy of the system $M–H+H_2O$.

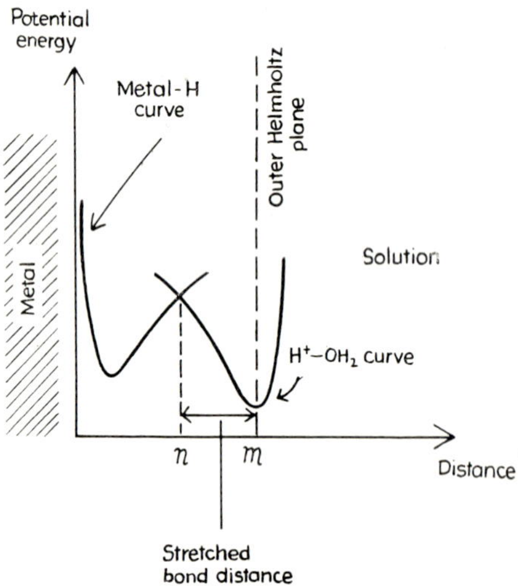

Fig. 4.2. When the hydrated proton is in its ground state (H^+ at the position m), the electron can't tunnel. When the $H^+–OH_2$ is in an activated state such that the H^+ is at the position n, the energy of an electron in the metal and one in H are equal, the tunnelling condition is satisfied, and the electron may tunnel from the metal to the proton.

Thus, one can represent the potential energy of the system $M(e)+H^+–OH_2$ as a function of the proton–water bond length, and the potential energy of the system $M–H+H_2O$ as a function of the position of the H atom between the metal and the water molecule. The corresponding curves showing how potential energy varies with distance are shown in fig. 4.2.

As the proton–water bond stretches, the potential energy of the system $M(e)+H^+–OH_2$ increases and when the H^+ arrives at the position n, the potential energies of the initial and final state intersect. Here, the tunnelling condition is satisfied, and the electron may tunnel to the proton. It is the existence of excited states corresponding to different stretched-bond positions of H^+ on the potential energy curve which permits an electron-energy state, of the same value E_F as the energy of the electron in the metal, to be available in the acceptor $H^+–OH_2$, so that radiationless tunnelling can occur.

In conclusion, electron transfer implies *co-operative movements of both proton and electron*, and these movements are influenced by the bond of the proton within the water molecule, and by the bond of the H atom to the metal ; the electron has to tunnel through a barrier like that in fig. 4.1 (*b*), but the activating of the proton–water bond is a prerequisite for this tunnelling.

This model brings out the role of $\Delta H_{hyd.\,H^+}$, i.e. of the heat of solvation between a proton and the surrounding water, as an important factor in determining the rate of the charge-transfer reactions at electrodes : it is $\Delta H_{hyd.\,H^+}$ which is involved in the change of energy in the different states of $H^+–OH_2$, and thus the system $M(e)+H^+–OH_2$ can reach the energetic situation in which the condition for tunnelling of an electron from the metal to the proton is satisfied.

It is now easy to understand why a potential difference, V, across the double layer affects an electrochemical reaction rate.

Consider fig. 4.3, where, for simplicity, the potential energy–distance relations have been approximated by straight lines.

Suppose first that there is no potential difference across the double layer. When the hydrated proton is in its ground state (at the outer Helmholtz plane, or in terms of the figure the H^+ is at position m), tunnelling cannot occur. To make tunnelling possible, ΔE_-^0 must be reduced to zero ; this can be done by increasing the energy of the system $M(e)+H^+–OH_2$ as a consequence of the proton–water bond stretching. The required activation energy is $\beta\Delta E_-^0$, where β is a positive number between 0 and 1 (generally about $\frac{1}{2}$) called ‘ the symmetry factor ’, and is of fundamental importance in charge transfer theory.

Suppose now that there is a negative potential difference, V, across the double layer. The most important effect will be the change of the energy of the electrons in the metal from E_F to $E_F - eV$ (the minus sign appears because the charge on the electron is negative). This change will *raise* the energy–distance curve of the system $M(e)+H^+–OH_2$ (because $-(eV)$ is positive), and thus give a *smaller* value to the energy difference. The value of this difference

$$\Delta E_- = \Delta E_-^0 + eV \qquad (4.5)$$

must be reduced to zero to make tunnelling possible, and this demands a smaller activating (or stretching) energy $\beta\Delta E_{-}$. Thus, a negative potential difference across the double layer increases the probability that H^{+}–OH_2 will exist in a quantum state suitable to fulfil the tunnelling condition. Hence, making the electrode potential more negative makes the electron ejection rate, or the cathodic current density, higher.

The anodic (de-electronating) reaction $M\text{–}H + H_2O \rightarrow H^{+}\text{–}OH_2$ can be analysed in the same way. It can be shown that a change of the electrode potential in the positive direction makes the rate of donation of electrons from H atoms to the metal faster.

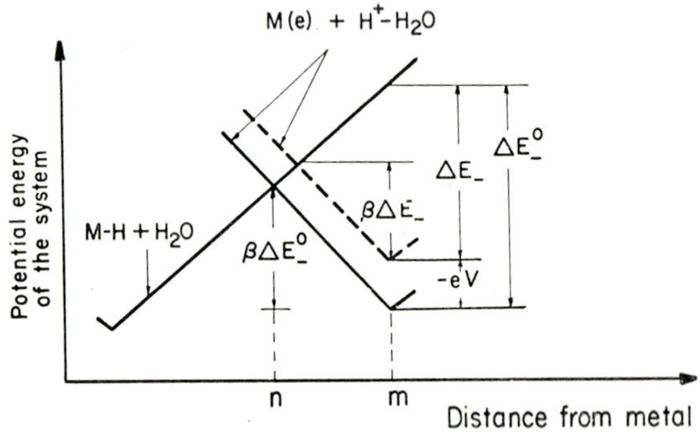

Fig. 4.3. The effect of a negative potential difference, V, across the double layer, on the cathodic (electronating) reaction $M(e) + H^{+}\text{–}OH_2 \rightarrow M\text{–}H + H_2O$.

4.6. *The relation between an electrochemical rate and the potential of the electrode at which it occurs*

Consider once again the reaction :

$$M(e) + H^{+}\text{–}OH_2 \rightarrow M\text{–}H + H_2O$$

and suppose that $V=0$, i.e. there is no charge on the electrode ; this is at *the potential of zero charge* (see Chapter 3, page 46). Then, any electron transfer would have to occur, like a chemical reaction, without any push from a potential difference.

The cathodic current density is given by :

$$(i_{-})_{V=0} = e n_{E_F} P_T n_{H_3O^+}{}^{*} \qquad (4.6)$$

where n_{E_F} is the number of electrons with Fermi energy E_F that collide per second with a unit area of the metal–solution interface, P_T is the tunnelling probability, and $n_{H_3O^+}{}^*$ is the number of H_3O^+ ions populating a unit area of the outer Helmholtz plane (OHP) and being suitably stretched to permit electron tunnelling.

P_T can be expressed as (equation (4.2)):

$$P_T \propto \exp\{-4\pi l h^{-1}[2m(E_B - E_F)]^{1/2}\} \tag{4.7}$$

where $E_B - E_F$ is the height of the barrier through which the electron has to tunnel.

For $n_{H_3O^+}{}^*$ one can use a Boltzmann distribution law:

$$n_{H_3O^+}{}^* = n_{H_3O^+} \exp(-\beta\Delta E\_{}^0/kT) \tag{4.8}$$

where $n_{H_3O^+}$ is the number of H_3O^+ ions populating a unit area of the OHP, and $\beta\Delta E\_{}^0$ is the stretching or activating energy (fig. 4.3) required to make tunnelling possible.

Thus,

$$(i\_)_{V=0} \propto e n_{E_F} \exp\{-4\pi l h^{-1}[2m(E_B - E_F)]^{1/2}\} \times \exp(-\beta\Delta E\_{}^0/kT) \times n_{H_3O^+}. \tag{4.9}$$

Consider now that there is a potential difference, V, across the double layer. Then $i\_$ is given by an equation similar to (4.9), except that the activation energy $\beta\Delta E\_{}^0$ is replaced (cf. equation (4.5)) by:

$$\beta\Delta E\_ = \beta\Delta E\_{}^0 + \beta e V$$

and therefore†:

$$i\_ = (i\_)_{V=0} \exp(-\beta eV/kT) = (i\_)_{V=0} \exp(-\beta FV/RT). \tag{4.10}$$

By a similar treatment one obtains the following expression for the anodic current density:

$$i_+ = (i_+)_{V=0} \exp[(1-\beta)FV/RT]. \tag{4.10'}$$

† We have cheated a bit here by writing the equation as though $E_B - E_F$ were independent of the potential of the electrode, V. This is not so, because the barrier through which the electrons have to tunnel varies in height with the electrode potential. However, the experimental facts fit equations (4.10) and (4.10′), and the reason why our neglect of the variation $E_B - E_F$ with potential works is that the exponential factor in which $E_B - E_F$ appears is small and the value of the exponential near to one.

4.7. *There are always two currents at an electrode–solution interface*

When we make the electrode more negative, electrons are expelled to the solution ; and when we make it more positive, they are attracted. However, at *any* condition of the electrode, there is some emission of electrons and some acceptance of them. It depends upon whether the electron-ejecting or the electron-accepting current is greater. We should therefore write a net electrochemical reaction rate as :

$$i_{\text{net, cathode}} = i_{-} - i_{+}. \tag{4.11}$$

This equation means : ' the net cathode current density observed in an outside circuit is the difference of the cathodic (electron emitting) current density and the anodic (electron accepting) current density '.

There must be an analogous equation for the net anode current (when the electrode is positive enough to make the partial anodic current predominate), and this will be :

$$i_{\text{net, anode}} = i_{+} - i_{-}. \tag{4.11'}$$

This equation would read : ' the net anode current density is the difference of the actual anodic current density, and the corresponding cathodic current density '.

Figure 4.4 illustrates the physical meaning of these two current densities. When the cathodic current is greater than the anodic current a net cathodic current, corresponding to $M^{+} + e \rightarrow M(e)$, flows through the interface; when the anodic current predominates a net anodic current, corresponding to $M(e) \rightarrow M^{+} + e$, flows through the interface.

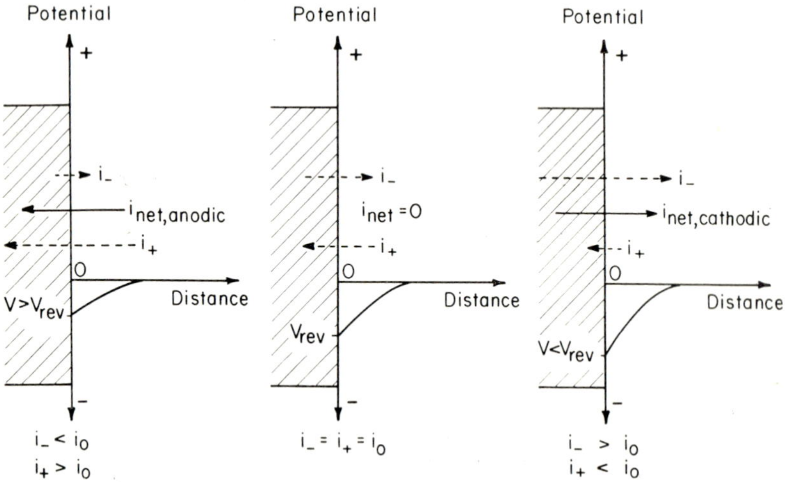

Fig. 4.4. The meaning of the exchange current density, i_0. The arrows indicate the electron flows through the interface. The reversible potential difference across the double layer is supposed to be negative.

4.8. *A relation between the net reaction rate and the electrode potential*

If one considers a net anodic current density as a positive quantity and a net cathodic current density as a negative quantity, one can use the following equation instead of the two equations (4.11) and (4.11′):

$$i_{net} = i_+ - i_-. \tag{4.12}$$

When $i_+ > i_-$, $i_{net} = i_{net,\ anodic} > 0$ and when $i_+ < i_-$, $i_{net} = i_{net,\ cathodic} < 0$.

Using the equations (4.10′) and (4.10) to express i_+ and i_- in terms of the potential difference, V, across the double layer, yields the following relation between the net density and the potential difference V:

$$i_{net} = (i_+)_{V=0} \exp [(1-\beta)FV/RT] - (i_-)_{V=0} \exp (-\beta FV/RT). \tag{4.13}$$

4.9. *The Butler–Volmer equation*

The relationship which we have deduced is that between the current and the potential of an electrode. One basic aspect of the relation has not yet been brought out. If there is a net anodic current when the electrode is sufficiently positive and a net cathodic current when the electrode is sufficiently negative, then there must exist a potential at which there is no net current, i.e. equilibrium in the electrode reaction would exist. Hence:

$$i_{net} = 0 \quad \text{when} \quad i_- = i_+. \tag{4.14}$$

This means that the rates of electron ejection (emission) and injection (acceptance) would be equal and the current zero.

This special potential at which the rates are equal is called by several names—the equilibrium potential, the reversible potential, or the thermodynamic potential. From now onwards, we shall refer to it as V_{rev}. Thus,

$$i_{net} = 0 \quad \text{when} \quad V = V_{rev}, \tag{4.14′}$$

and from equation (4.13) it follows that:

$$i_0 = (i_+)_{V=0} \exp [(1-\beta)FV_{rev}/RT] = (i_-)_{V=0} \exp (-\beta FV_{rev}/RT) \tag{4.15}$$

We call i_0 the *exchange current density.*

Figure 4.4 illustrates also the meaning of exchange current density: at equilibrium (i.e., when $V = V_{rev}$), the partial anodic and cathodic current densities (positive equilibrium) are equal, and $i_0 = (i_+)_{rev} + (i_-)_{rev}$. When V is more positive than V_{rev}, $i_- < i_0$, $i_+ > i_0$, and a net anodic current (electrons from solution to metal) flows through the interface. When V is more negative than V_{rev}, $i_- > i_0$, $i_+ < i_0$, and a net cathodic current flows through the interface.

Equation (4.13) can be written in terms of V_{rev}:

$$i=(i_+)_{V=0}\exp[(1-\beta)FV_{rev}/RT]\exp[(1-\beta)F(V-V_{rev})/RT]$$
$$-(i_-)_{V=0}\exp(-\beta FV_{rev}/RT)\exp(-\beta F(V-V_{rev})/RT). \quad (4.13')$$

Then, using the expressions of i_0 (equation (4.15)):

$$i=i_0\{\exp[(1-\beta)F(V-V_{rev})/RT]-\exp(-\beta F(V-V_{rev})/RT)\}. \quad (4.16)$$

The difference

$$\eta=V-V_{rev} \quad (4.17)$$

expresses *how far away from the equilibrium* (*or reversible*) *potential* we are, and is called the *overpotential* of the electrode reaction.

The overpotential η is the *extra* potential that must be applied to an electrode reaction to make it occur at a certain net velocity. When $\eta=0$, the electrode is at its reversible position and no *net* current passes. The cathodic and anodic *partial* reactions *are still occurring*, but $i_{net}=0$ (i.e. $(i_-)_{rev}=(i_+)_{rev}=i_0$). If $\eta>0$ (i.e. $V>V_{rev}$), $i_-<i_0$ and $i_+>i_0$; a net anodic current flows through the interface. If $\eta<0$ (i.e. $V<V_{rev}$), $i_->i_0$ and $i_+<i_0$; a net cathodic current flows through the interface (see also fig. 4.4).

Now, we can go back to equation (4.16) and replace $V-V_{rev}$ by η:

$$i=i_0\{\exp[(1-\beta)F\eta/RT]-\exp[-\beta F\eta/RT]\}. \quad (4.18)$$

This is the Butler–Volmer equation, and relates the net current density of the electrochemical reaction to the overpotential. The Butler–Volmer equation shows that:

At positive (*anodic*) overpotentials, i is positive, so that *the net current is anodic*, and

At negative (*cathodic*) overpotentials, i is negative, so that *the net current is cathodic*.

As we put more electrons into our electrode, or take more out, we are altering the level of the Fermi electrons. Hence, the change in the Fermi level in the metal from the value it had at equilibrium measures the overpotential, and both correspond to the net reaction rate (or current density) to which they give rise.

4.10. *Special cases of the Butler–Volmer equation*

4.10.1. *The Tafel equation*

Let us suppose that the electrochemical reaction occurs at large anodic overpotentials, i.e. at $\eta \gg RT/F$.

It can be seen from the form of the equation that the second term in equation (4.18) can be neglected, so that:

$$i = i_0 \exp[(1-\beta)F\eta/RT]. \tag{4.19}$$

If the electrochemical reaction occurs at large cathodic overpotentials, i.e. at $\eta \ll RT/F$, it is the first term in equation (4.18) that can be neglected, and the Butler–Volmer equation reduces to :

$$i = -i_0 \exp(-\beta F\eta/RT). \tag{4.19'}$$

Equations (4.19) and (4.19′) can be rewritten as :

$$\eta = -\frac{RT}{(1-\beta)F}\ln i_0 + \frac{RT}{(1-\beta)F}\ln i\,; \quad (\eta \gg RT/F) \tag{4.20}$$

$$-\eta = -\frac{RT}{\beta F}\ln i_0 + \frac{RT}{\beta F}\ln(-i)\,; \quad (\eta \ll -RT/F). \tag{4.20'}$$

Equations (4.20) and (4.20′) are known as *Tafel equations.*

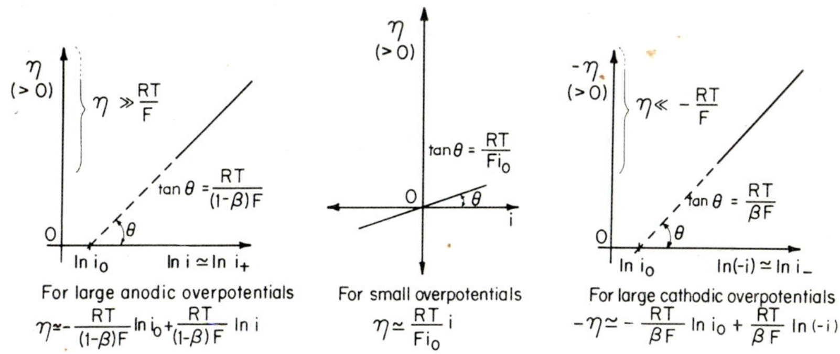

Fig. 4.5. Three particular cases of the Butler–Volmer equation. The Tafel equation for large anodic overpotentials ($\eta \gg RT/F$) ; the linear relation for small anodic or cathodic overpotentials $|\eta| \ll RT/F$; and the Tafel equation for large cathodic overpotentials ($\eta \ll -RT/F$).

Consider now an experiment performed at large *anodic overpotentials.* Equation (4.20) shows that the plot η against $\ln i$ should be a straight line with the slope given by $RT/(1-\beta)F$ (see also fig. 4.5 (*a*)) ; RT/F is a known quantity, and β can be derived from the experimental value of this slope.

Equation (4.20) also shows that η is zero for $\ln i = \ln i_0$; extrapolation to $\eta = 0$ (the dotted line in fig. 4.5 (*a*)), gives the value of i_0.

Figure (4.5 (*c*)) shows how one can obtain the values of β and i_0 from experiments performed at large cathodic overpotentials, when equation (4.20′) applies.

4.10.2. *The linear relation between i and* η

Consider very small anodic or cathodic overpotentials, i.e. $|\eta| \ll -RT/F$. Then, the exponentials in the Butler–Volmer equation can be expanded, and taking the first term only in $e^x \simeq 1 + x + \ldots$ valid for small values of x, give the following linear relationship :

$$i \simeq i_0[1+(1-\beta)F\eta/RT-(1-\beta F\eta/RT)] = \left[\frac{i_0F}{RT}\right]\eta. \qquad (4.21)$$

This can be put in the more convenient form :

$$\eta \simeq \frac{RT}{F}\frac{i}{i_0}. \qquad (4.21')$$

Thus, in the range of very small overpotentials, < 10 mV, say, the plot of η against $\ln i$ is a straight line, and from its slope RT/Fi_0 one can obtain the value of i_0 (see also fig. 4.5 (*b*)).

4.11. *Electrochemical reactions*

The Butler–Volmer equation is important, for much of basic electrochemistry can be described in terms of it.

We have shown (page 64) that we can readily change the velocity of electrochemical reactions by pushing in or taking out tiny amounts of excess charge on the surface of the electrode. Consider an electrochemical reaction taking place at a large anodic overpotential, i.e. when equation (4.19) applies, and let us change the overpotential by 0·20 V. Then, the current density (representing the reaction velocity) would change by (equation (4.19)) :

$$i_{\eta+0\cdot2}/i_\eta = \exp\,[(1-\beta)F\,0\cdot2/RT]. \qquad (4.22)$$

Taking $F = 96\,500$ C mol^{-1}, $R = 8{\cdot}317$ J K^{-1} mol^{-1}, $T = 298$K, and $\beta = 0{\cdot}5$:

$$i_{\eta+0\cdot2}/i_\eta = \exp\,(0{\cdot}5 \times 0{\cdot}2 \times 96\,500/8{\cdot}317 \times 298) \simeq 100.$$

Thus, the current density is *increased* about a hundredfold when the potential is made more positive by 0·2 V.

4.12. *How to measure overpotential*

In traditional chemistry, reaction rates are often measured as a function of the temperature. The corresponding measurement in electrochemistry is that of the current (representing the reaction rate) as a function of *overpotential* (equations (4.19) and (4.19′)).

It is easy to measure a current, but how does one measure an overpotential ? The method of doing it is shown in fig. 4.6.

By coupling the interface under study (which we call ‘ the working electrode ’) with a counter-electrode via a *polarizing* circuit, a control-

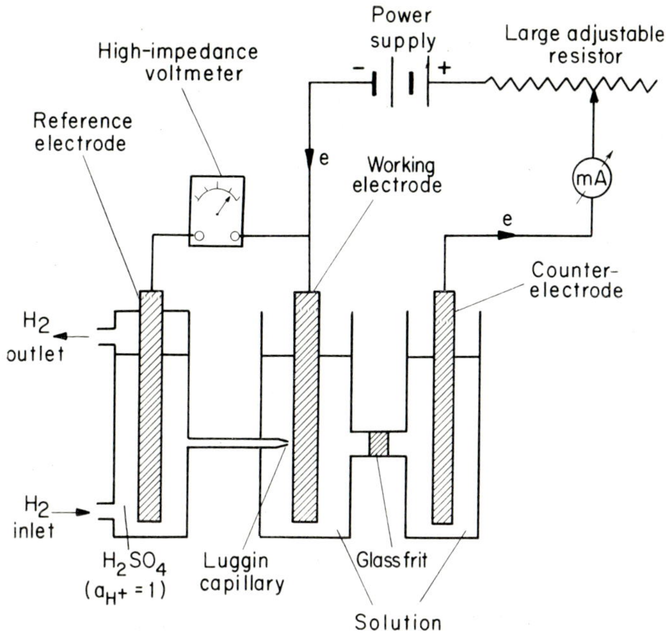

Fig. 4.6. The principle of the three-electrode system required to measure electrode overpotentials.

lable current is made to pass and to produce changes of the potential of the working electrode. As the overpotential is the difference between the potential which the electrode has for a certain net current density, i, and the potential which it has for zero i (the reversible potential), the change in potential caused by passing a current is the quantity that we wish to measure.

Such changes are measured by coupling the working electrode with a reference electrode. This reference electrode behaves reversibly with respect to the reaction at the interface; no net current passes across it and hence it gives the potential at equilibrium.

Suppose now that one uses as reference an electrode identical in composition with the working electrode. The voltmeter should then read zero when no current flows through the *polarizing* circuit, because both the working electrode and the reference electrode are at the same equilibrium potential, V_{rev}.

When a current of current density i passes through the polarizing circuit, the voltmeter reading gives the potential of the working electrode relative to the potential, V_{rev}, of the reference electrode. This is the overpotential, $\eta = V - V_{rev}$ (equation (4.17)). This value of η

that is measured is just for that particular current density and as i changes, η will also change.

4.13. *The rate determining step*

We have discussed only the simplest electrochemical reactions, like:

$$M^+ + e \rightarrow M$$

$$H_3O^+ + e \rightarrow H_{ads} + H_2O.$$

However, most reactions in electrochemistry are more complicated than these. Let us consider the further reactions involved in hydrogen *evolution*, not merely discharge of H_3O^+ to H atoms adsorbed on an electrode. There are two steps. The first is:

$$H^+ + e \rightarrow H_{ads}. \qquad (4.23)$$

The second may sometimes be:

$$2H_{ads} \rightarrow H_2. \qquad (4.24)$$

The important question is, which of these steps is the *rate determining* step of the reaction $2H^+ + 2e \rightarrow H_2$?

Let us suppose that the rate constant for the discharge of protons to form adsorbed hydrogen atoms is very much smaller than that for the rate of combination of adsorbed hydrogen atoms.

We can then say that the formation of H_2 molecules from H_{ads} atoms (once they arrive through the double layer at the surface) is relatively instantaneous. Consequently, the rate of the overall reaction $2H^+ + 2e \rightarrow H_2$ is largely determined by the rate of the electronation reaction $H^+ + e \rightarrow H_{ads}$, i.e. *the first step* (4.23) *is rate determining.*

In this case, for a large cathodic overpotential, we can use equation (4.20′) to express the relation between η and i:

$$-\eta = -\frac{RT}{\beta F}\ln i_0 + \frac{RT}{\beta F}\ln(-i). \qquad (4.20')$$

Therefore, if $\beta = 0{\cdot}5$ and the first step (4.23) is the rate determining one, the gradient of the linear part of the experimental curve of $-\eta$ against $\ln(-i)$ should be $2RT/F$.

Suppose, on the other hand, that the rate of the combination reaction (4.24) is smaller than the rate of the electronation reaction (4.23).

Then, hydrogen atoms are easily available by means of the electronation reaction (now assumed to be relatively rapid), but they find difficulty in getting *off* the electrode surface by means of the combination reaction. The rate of the overall production of H_2 will be determined by the rate of combination i.e. *the second step* (4.24) *is rate determining.*

The combination reaction is a chemical reaction and its rate is proportional to the square of the concentration of H_{ads}:

$$v = k[\mathrm{H_{ads}}]^2. \tag{4.25}$$

To find the relation between η and i, we have to relate $[\mathrm{H_{ads}}]$ to the overpotential, η.

Before doing this, let us return to the electronation reaction and denote by v_- the rate of reaction $\mathrm{H^+ + e \rightarrow H_{ads}}$, and by v_+ the rate of the reaction $\mathrm{H_{ads} \rightarrow H^+ + e}$. Then $v_- - v_+$ expresses the net rate of formation of hydrogen atoms ; when the steady state is reached, this net rate must equal the rate, v, of transformation of $\mathrm{H_{ads}}$ atoms to H_2 molecules. Because $v \ll v_-$, it follows that, in a relative sense, $v_- - v_+ \sim 0$, i.e. $v_- \simeq v_+$ (so that the forward reaction (+) and the backward reaction (−) are *almost* in equilibrium).

Hence, when it is not the rate-determining step in the sequence, the proton discharge reaction to form adsorbed H atoms may be treated as being in pseudo-equilibrium.

We can therefore apply the equation which relates an equilibrium (' reversible ') potential to the concentration of atoms on the surface and the concentration of protons in solution, because our situation is (almost) an equilibrium one. This equation is the Nernst equation (see Chapter 5) :

$$V = V^\circ + \frac{RT}{F} \ln \frac{c_{\mathrm{H^+}}}{[\mathrm{H_{ads}}]}. \tag{4.26}$$

But $V = V_{\mathrm{rev}} + \eta$ (equation (4.17)). Introducing this in equation (4.26) and taking the antilogarithm :

$$[\mathrm{H_{ads}}] = K \exp(-F\eta/RT) \tag{4.27}$$

where V_{rev}, V°, and $c_{\mathrm{H^+}}$, being constant, were included in the K.

Hence, equation (4.25) becomes :

$$v = kK^2 \exp(-2F\eta/RT). \tag{4.28}$$

If now we use equation (4.3) to express v in terms of i :

$$i = -2FkK^2 \exp(-2F\eta/RT) = -K_1 \exp(-2F\eta/RT) \tag{4.29}$$

where the minus sign takes into account the fact that the net current is cathodic. Equation (4.29) can also be presented in logarithmic form :

$$-\eta = -\frac{RT}{2F} \ln K_1 + \frac{RT}{2F} \ln(-i). \tag{4.30}$$

If we compare with equation (4.20′), we see that now the gradient is $RT/2F$, *four times smaller* than the gradient expected when the current is rate controlled by the discharge of protons.

Thus in the sequence $\mathrm{2H^+ + 2e \rightarrow 2H_{ads}}$; $\mathrm{2H_{ads} \rightarrow H_2}$, where electron transfer is rate-determining, there will be one kind of relation between n $(-i)$ and η (gradient $= 2RT/F$); whereas when the combination of atoms is rate determining, the relation between ln $(-i)$ and η will be different (gradient $= RT/2F$).

Rate-determining steps can occasionally be established by examining the difference in the gradient of the ln i/η relation in this way. It is rare that matters are so simple, particularly as several mechanisms may give the same gradient. More often one has to determine several different characteristics of the reaction (for example, how the degree of coverage of the electrode surface with reaction atoms varies with potential), and argue which of the possible rate-determining steps is the most consistent with the results of the various (diagnostic) measurements.

4.14. *Short time measurements in the kinetics of electrode reactions: transient or sweep measurements*

It is often necessary in electrochemistry to make measurements over times of the order of milliseconds, sometimes microseconds, and occasionally nano-seconds. One reason for this is that quick measurement means less chance of the reaction surface—the *electrocatalytic* surface—changing during measurement, which would lead to difficulties in interpretation.

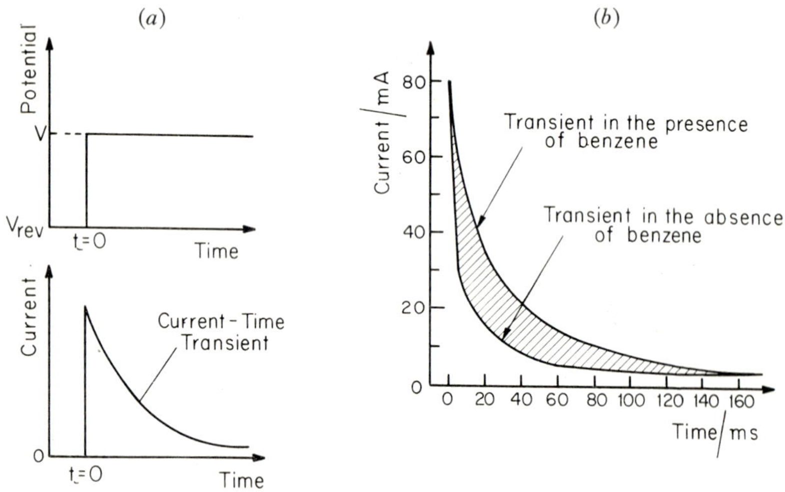

Fig. 4.7. (*a*) The time dependence of the *stimulus* (potential step of height $V - V_{rev}$), and of the *response* (current-time transient) in the potentiostatic transient method. (*b*) Current-time transients—in the presence and in the absence of benzene—when a potential step is applied to a platinized platinum electrode in 0·5 M H_2SO_4. The shaded area expresses the charge required to oxidize the adsorbed benzene to CO_2.

Transient measurements sometimes yield information on the constitution of substances on the surface of electrodes. For example, we can obtain information about how much benzene is adsorbed from solution on an electrode by making a transient measurement, at constant potential, the benzene being oxidized to CO_2. During the short interval concerned, the electrode is maintained at a constant potential, as represented by a step function (fig. 4.7 (*a*)), and the current is measured as a function of time.

The electrode is first held at a known potential to reach adsorption equilibrium (fig. 4.7 (*b*)), and then its potential is changed to a region in which organic oxidation occurs. The area under the current–time transient in the presence of benzene expresses the charge (because charge $= \int_0^t \text{current} \times \mathrm{d}t$) required to oxidize the adsorbed benzene, plus the charge required for oxidizing the substrate metal.

To obtain the charge required to oxidize only the adsorbed benzene, we record also the current–time transient in the absence of benzene. The area under this second current–time transient represents the charge required for the oxide formation of the substrate metal, and thus the area between the two transients gives the charge required to oxidize the adsorbed benzene. Let Q be this charge.

Assuming complete oxidation according to the reaction :

$$C_6H_6 + 12H_2O \rightarrow 6CO_2 + 30H^+ + 30e$$

it is seen that the oxidation of each mole of benzene requires 30 moles of electrons, that is, a charge of 30 F. Thus, the number of moles of adsorbed benzene present on the electrode is given by $Q/30F$.

Most modern electrochemistry is carried out by *transient measurements*, which are simpler to interpret than measurements carried out over longer times with constant current, because the surface changes negligibly during the measurement. One can thus obtain information about free radicals on the electrode.

4.15. *Transport control*

We have talked about the relationship between the rates of an electrochemical reaction and the potential across the double layer as if there were no limit to the rate at which an electrochemical reaction could take place. Equation (4.18), suggests that by increasing the overpotential further and further, one should be able to get up to any current.

Common-sense indicates that this cannot be so. One reason is that we have neglected to consider the supply of ions or molecules to the electrochemical reaction zone, viz. the double layer, from the bulk of the solution.

If diffusion, convection, stirring, etc., cannot supply the ions fast enough, then the current will be limited by one of these processes which transport the ions to the interface. It will not follow the Butler–Volmer equation (equations (4.18), (4.20) and (4.20′)). During the discussion of that equation, it was assumed that there were no limitations due to *transport*; only the *charge-transfer* limitations were taken into account.

Consider the case in which the charge-transfer limitations are small or negligible compared to the transport limitations.

This would happen if the charge-transfer overpotential (the type dealt with in the Butler–Volmer equation) is negligible compared to the overpotential produced by a concentration change at the interface. Thus, if transport of ions to and from the interface is easy, the concentration at the interface is virtually that obtaining in the bulk. But if transport is relatively poor, the ionic concentration at an interface between metal and solution will be less than that obtaining in the bulk. The transport of ions to the electrode will not have kept up with the demand of the electron flow. The overpotential caused by this concentration change is known as the *concentration overpotential*, η_c.

Now, when one can neglect the charge-transfer overpotential? The Butler–Volmer equation shows that when η tends to zero, i_0 must tend to infinity for i to remain finite; one can neglect the charge-transfer overpotential by comparison with the concentration overpotential when the exchange-current density of the charge transfer reaction is sufficiently great.

Table 4.1 gives some values of i_0 for the reaction of hydrogen evolution, on various metals.

Metal	i_0/A m^{-2}
Hg	$10^{-8.2}$
Cd	$10^{-6.9}$
Ti	$10^{-4.2}$
Nb	$10^{-3.0}$
W	$10^{-2.0}$
Au	$10^{-1.5}$
Ni	$10^{-1.2}$
Rh, Ir	$10^{+0.3}$
Pd, Pt	$10^{+0.9}$

Table 4.1. Approximate exchange current densities for the hydrogen-evolution reaction on various metals in M H_2SO_4.

Consider such a situation in which the charge-transfer overpotential can be neglected. The charge transfer reaction is assumed to be near equilibrium, and one can therefore use the Nernst equation to express the potential difference across the interface.

Thus, when the current is zero :

$$V_{\text{rev}} = V_{\text{rev}}^{\circ} + \frac{RT}{F} \ln c_{\text{bulk}} \tag{4.31}$$

because the concentration at the electrode surface is equal to the bulk concentration.

When a current density i passes through the interface, $c_e \neq c_{\text{bulk}}$, and :

$$V = V_{\text{rev}}^{\circ} + \frac{RT}{F} \ln c_e. \tag{4.32}$$

Thus :

$$\eta_c = V - V_{\text{rev}} = \frac{RT}{F} \ln (c_e/c_{\text{bulk}}). \tag{4.33}$$

In Chapter 2, we have shown that the rate (expressed as a current density) of passage of material down a concentration gradient is given by :

$$i = -DF(\mathrm{d}c/\mathrm{d}x)$$

and we have used the Nernst model to express the concentration gradient. Diffusion occurs across a *Nernst diffusion layer* of constant thickness δ, and $\mathrm{d}c/\mathrm{d}x = (c_{\text{bulk}} - c_e)/\delta$). So, from equation (2.20) :

$$i = -DF \frac{c_{\text{bulk}} - c_e}{\delta}. \tag{4.34}$$

When c_e becomes zero, i has its greatest value, *the limiting current density* i_L :

$$i_L = -DF(c_{\text{bulk}}/\delta). \tag{4.34'}$$

From equations (4.34) and (4.34′) it follows that

$$c_e/c_{\text{bulk}} = 1 - (i/i_L) \tag{4.35}$$

and thus :

$$\eta_c = \frac{RT}{F} \ln \left(1 - \frac{i}{i_L}\right) \tag{4.36}$$

or :

$$i = i_L[1 - \exp (F\eta_c/RT)]. \tag{4.36'}$$

Equation (4.36′) gives the relation between the current density i and the concentration overpotential when the electrochemical reaction is restricted only by transport limitations (fig. 4.8).

Equation (4.33) shows that (because $c_e < c_{\text{bulk}}$) $\eta_c < 0$, consequently, i_L is a *cathodic limiting current density.*

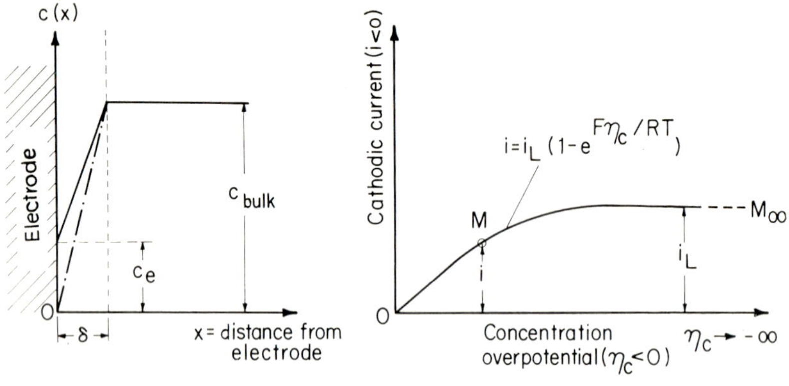

Fig. 4.8. The Nernst diffusion layer of constant thickness δ; and the relation between i and η_c, when only the concentration overpotential η_c is taken into account. As c_e tends to zero, the point M on the curve i vs. η_c tends to the point M_∞.

In an unstirred solution, the value of δ is of the order of 5×10^{-4} m. In a stirred solution it may be decreased with rapid stirring to $< 10^{-6}$ m, so i_L may be *increased* some 500 times by stirring.

When can we use the Butler–Volmer equation and then equation (4.36′) to relate i and η? A numerical solution of equation (4.36′) shows that, if $i/i_L < 0{\cdot}1$, $\eta_c < 0{\cdot}003$ V. This is a small overpotential so that, granted the condition $i < 0{\cdot}1\, i_L$, we can usually neglect the concentration overpotential and apply the Butler–Volmer equation (4.18) (and the useful limiting cases, 4.20 and 4.20′) alone. Now, if $i/i_L > 0{\cdot}5$, say, then $|\eta_c|$ will be $> 0{\cdot}15$ V, which is significant; and equation (4.36′) begins to be a better way than the Butler–Volmer equation to represent the relation of current to overpotential.

A more complex equation, taking into account both the shift of the Fermi level due to electron transfer, and the shift of the equilibrium potential due to the change of interfacial potential which is associated with diffusion difficulties, can be worked out. But it is only of practical use over a small range of overpotentials and is also more complex.

CHAPTER 5

electrochemical science in chemistry

5.1. *Introduction*

SEVERAL basic and elementary aspects of physical chemistry originated in electrochemistry. For example, the concepts of pH, dissociation constants, and solubility product arose from electrochemistry and have become the basis of analytical chemistry.

Colloid chemistry is the study of particles in the micrometre (μm) size region. Here the repulsion between the electrical charges of like sign at the interface prevents the particles coagulating and therefore preserves their stability.

The third law of thermodynamics (the entropy of an ideal crystalline solid is zero at the absolute zero of temperature) originated in studies of galvanic cells, where it was found that at low temperatures the entropy of a reaction in an electrochemical cell tended to zero as the temperature decreased.

In our discussion, we shall deal with the relation between the energy change in a chemical reaction and the potential difference in a cell; the electrochemical aspects of catalysis; and electrochemical syntheses.

5.2. *Energetics*

5.2.1. *The Nernst equation*

Let us go back to equation (4.15).

$$(i_+)_{V=0} \exp\,[(1-\beta)FV_{\text{rev}}/RT] = (i_-)_{V=0} \exp\,(-\beta FV_{\text{rev}}/RT) \quad (5.1)$$

or:

$$V_{\text{rev}} = \frac{RT}{F} \ln \frac{(i_-)_{V=0}}{(i_+)_{V=0}}. \quad (5.2)$$

In an aqueous solution, the number n_{M^+} of M^+ ions in contact with the electrode is proportional to the concentration of M^+ ions in the bulk of the solution, and thus equation (4.9) can be written:

$$(i_-)_{V=0} = k_- \exp\,(-E_{(-)}/kT)c_{M^+} \quad (5.3)$$

where k_- is independent of concentration, and $E_{(-)}$ is an activation energy for the reaction $M^+ + e \rightarrow M$.

A similar equation for $(i_+)_{V=0}$ is:

$$(i_+)_{V=0} = k_+ \exp\,[E_{(+)}/kT] \quad (5.3')$$

where $E_{(+)}$ is the energy of activation for the reaction $M \rightarrow M^+ + e$. There is no corresponding concentration term because, for the dissolution of the metal, the whole surface is taken as the source of dissolving ions. The ' concentration ' of atoms per unit area is therefore almost independent of the concentration of ions in the bulk.

Thus, equation (5.2) becomes :

$$V_{\text{rev}} = V_{\text{rev}}^{\circ} + \frac{RT}{F} \ln c_{\text{M}^+} \tag{5.4}$$

where :

$$V_{\text{rev}}^{\circ} = \frac{RT}{F} \ln \frac{k_-}{k_+} + \frac{N_{\text{A}}}{F} [E_{(+)} - E_{(-)}]. \tag{5.4'}$$

V_{rev}° is called the *standard reversible potential difference* and represents a particular value of V_{rev}, i.e. the value of V_{rev} when c_{M^+} is *standardized to unity*.

For a more general case of an electron acceptor A^{z+} in equilibrium with an electron donor D (i.e. $D \rightleftharpoons A^{z+} + ze$) :

$$V_{\text{rev}} = V_{\text{rev}}^{\circ} + \frac{RT}{zF} \ln \frac{a_{\text{A}^{z+}}}{a_{\text{D}}} \tag{5.5}$$

where $a_{\text{A}^{z+}}$, a_{D} are the activities (see Chapter 2, page 22), of the electron acceptor and electron donor respectively.

Equation (5.5), first derived by Nernst, tells how the potential of an interface varies with the activity (roughly, the concentration) of ions in solution. In dilute solutions, the activities can be replaced by concentrations.†

The Nernst equation can be applied only rarely to the study of electrodes in solution, because it is quite difficult to find interfaces which do exhibit the simple kind of equilibrium (i.e. just one reaction, such as $Ag^+ + e \rightleftharpoons Ag$) in control. More often, several reactions determine the behaviour of an electrode when no current passes across it. However, Nernst's equation is of immense basic significance because it tells us, in terms of basic thermodynamic quantities (standard free energies of the ions etc. which take part in the reaction) what the potential (V_{rev}) of an electrode *at equilibrium* ought to be (see equation (5.11)). It gives us the potential to which we refer the overpotential in kinetics (for $\eta = V - V_{\text{rev}}$).

† As an approximation we can replace activity by *concentration* in mol dm^{-3}. Electrical interactions between the ions in fact make the value of the activity (in the sense of ' reacting power ') different per ion at different concentrations. See p. 22.

5.2.2. *The self-driving cell* (an energy producer ; see Chapter 1, page 9). Consider for example a cell :

$$\begin{matrix} \text{Electrode (1)/Electrolyte/Electrode (2)} \\ \text{I} \qquad\qquad \text{S} \qquad\qquad \text{I}' \end{matrix} \tag{5.6}$$

and let the electrode reactions be :

$$\begin{aligned} D_1 &\rightarrow A_1^{z+} + ze \\ A_2^{z+} + ze &\rightarrow D_2. \end{aligned} \tag{5.6'}$$

Then, summing,

$$D_1 + A_2^{z+} \rightarrow A_1^{z+} + D_2 \tag{5.6''}$$

is the *chemical* overall reaction in the cell.

The difference in chemical potentials is :

$$\Delta G = \mu_{A_1z+} + \mu_{D_2} - \mu_{D_1} - \mu_{A_2z+}$$

where $\mu_i = \mu_i^\circ + RT \ln a_i$ is the chemical potential of the corresponding species i (see Chapter 3, page 34). It represents the *free energy* change in the chemical reaction (5.6″). Thus

$$\Delta G = \Delta G^\circ + RT \ln \frac{a_{A_1z+} a_{D_2}}{a_{D_1} a_{A_2z+}} \tag{5.7}$$

where

$$\Delta G^\circ = \mu^\circ{}_{A_1z+} + \mu^\circ{}_{D_2} - \mu^\circ{}_{D_1} - \mu^\circ{}_{A_2z+} \tag{5.7'}$$

ΔG° is the *standard* free energy change, i.e. the value of ΔG when $a_{A_1z+} a_{D_2} / a_{D_1} a_{A_2z+}$ is equal to 1.

Suppose now that the electrical resistance of the external load is very great so that the cell is passing a very small current. One can then say that the cell is operating at a nearly-equilibrium condition, because a negligible current across the electrodes means a negligible overpotential at the electrodes. Hence, one can apply the Nernst equation (5.5) to calculate the potential differences at the two interfaces. Thus :

$$\begin{aligned} V_{rev;\,I'} &= V^\circ{}_{rev;\,2} + \frac{RT}{zF} \ln (a_{A_2z+}/a_{D_2}) \\ V_{rev;\,I} &= V^\circ{}_{rev;\,1} + \frac{RT}{zF} \ln (a_{A_1z+}/a_{D_1}). \end{aligned} \tag{5.8}$$

But :

$$V_{rev;\,cell} = V_{rev;\,I'} - V_{rev;\,I} \tag{5.9}$$

is the *reversible* voltage of the cell, and therefore (from equations (5.8) :

$$zFV_{\text{rev; cell}} = zFV^{\circ}_{\text{rev; cell}} - RT \ln \frac{a_{A_1z+}a_{D_2}}{a_{D_1}a_{A_2z+}} \tag{5.10}$$

where :

$$V_{\text{rev; cell}} = V^{\circ}_{\text{rev; 2}} - V^{\circ}_{\text{rev; 1}} \tag{5.10'}$$

is called the *standard* reversible voltage of the cell, i.e. the value of $V_{\text{rev; cell}}$ when $a_{A_1z+}a_{D_2}/a_{D_1}a_{A_2z+} = 1$.

By comparing equations (5.7) and (5.10) :

$$zFV_{\text{rev; cell}} = -\Delta G. \tag{5.11}$$

This relation is the most fundamental one in the study of the energy relations of electrochemical cells and, therefore, is the basic equation in electrochemical energy conversion. It tells us that the potential developed by a cell at equilibrium depends only on the free energy change in the chemical reaction which occurs in the cell. No properties of the substrates come into the discussion at all. Thus, the free energy change in the reaction in the electrochemical cell is only that of the overall chemical reaction, and not a function of the energy of the electrons in the metals concerned (for no term for that energy appears in 5.7 or 5.10).

So, considering a hydrogen–oxygen cell operating with a negligible current flow (i.e. virtually at equilibrium), we should have :

$$2\,H_2 \rightarrow 4\,H^+ + 4\,e \qquad \text{(at the anode)}$$

$$O_2 + 4\,H^+ + 4\,e \rightarrow 2\,H_2O \qquad \text{(at the cathode)}$$

and the overall *chemical* reaction is :

$$2\,H_2 + O_2 \rightarrow 2\,H_2O.$$

This last equation is the cell reaction, and it will be seen that the energy of the electrons in the metal of the electrodes does not affect the thermodynamics of the reaction.

These relations between electrode potential and free energy allow the use of electrochemical measurements to find the free energy of chemical reactions by measuring the potential, E, of a cell in which the reaction occurs, and then applying equation (5.11). This is relatively easy to carry out with a number of reactions for which the i_0 (p. 63) is large, so that the reactions attain equilibrium easily without being disturbed by the effects of possible other reactions in the same solution with lower i_0 values and hence (equation (4.19)) less contribution to the current across the interface. Such measurements helped to lay the foundations of thermochemistry, which is concerned with the entropy and free energy of substances, the most basic of all chemical quantities.

5.3. *Electrocatalysis*

5.3.1. *What does electrocatalysis mean?*

We have seen from equation (4.18) that the velocity of an electrochemical reaction, measured in terms of the current (the number of electrons per unit time) which flows across a unit area of an electrode, is given by :

$$i = i_0\{\exp[(1-\beta)F\eta/RT] - \exp(-\beta F\eta/RT)\} \qquad (4.18)$$

and under many practical conditions by the formulae :

$$i = i_0 \exp[(1-\beta)F\eta/RT] \quad \text{(anodic reaction)†} \qquad (4.19)$$

$$i = -i_0 \exp(-\beta F\eta/RT) \quad \text{(cathodic reaction)†} \qquad (4.19')$$

We have shown in Chapter 4 that the current through an electrochemical system, and in consequence the reaction rate, can be varied through many orders of magnitude by changing the overpotential ; this can be done readily by changing the voltage applied to the system from an external power source.

One could refer to *this* as electrocatalysis. The meaning would be 'the increase of a reaction rate caused by the electrical potential difference (represented by the overpotential) ' in the double layer. However, this meaning is a trivial one and it is *not* what is usually meant by electrocatalysis.

Let us consider, for example, oxygen evolution ($2\,H_2O \rightarrow O_2 + 4\,H^+ + 4\,e$). It may take place on various metals, and at different overpotentials. Figure 5.1 shows a series of Tafel lines for oxygen evolution on various substrates in M $HClO_4$. These Tafel lines are described mathematically by the logarithmic form of equation (4.19) :

$$\eta = -\frac{2{\cdot}3\,RT}{(1-\beta)F}\log_{10} i_0 + \frac{2{\cdot}3\,RT}{(1-\beta)F}\log_{10} i \qquad (5.12)$$

and the intercepts with the axis for $\eta = 0$ give the values of $\log_{10} i_0$.

For a given overpotential, the difference in the rate of the reaction on various substrates is given by the difference in the values of the current densities, i. Then, fig. 5.1 shows :

$$(i)_{\mathrm{Pt}} < (i)_{\mathrm{Pt\text{-}Rh}} < (i)_{\mathrm{Rh}} < (i)_{\mathrm{Ir}}$$

and because the Tafel lines are practically parallel, the same conclusion remains valid for the values of the exchange current densities, i_0.

Now, all current densities, or rates, in electrode kinetics are proportional to rate constants (equations (5.3), (5.3')). Hence, i_0 is proportional to the rate constant and its variation (for a given electrode

† For these equations to be substantially valid, the numerical values of η must be > 50 mV (at room temperatures).

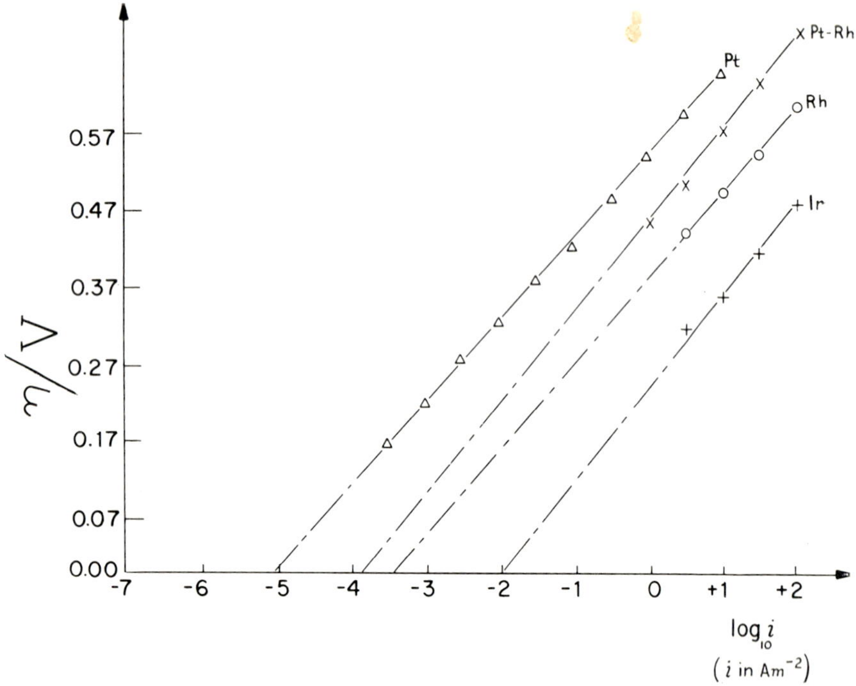

Fig. 5.1. η versus $\log_{10} i$ plots for pure Pt(Δ), Pt–Rh alloy (40atom-% Rh) (x), pure Rh(o) and Pure Ir(+).

reaction) with the nature of the electrode substrate constitutes the subject of electrocatalysis. It is really more complicated than that, because, as shown in equations (4.15), (5.3) and (5.3′), the value of i_0 depends not only on a *rate constant*, but also on the concentration of the reactants. However, i_0 may be regarded as the chief factor affecting the *rate* (and so the catalysis) at a given overpotential. The higher the value of i_0 for a given reaction, the better is the corresponding substrate as an electrocatalyst for that reaction ; that means the overpotential developed at a given rate will be smaller than on a less active catalyst (equation (5.12)). For example, the Tafel lines in fig. (5.1) indicate that for the oxygen evolution reaction from M $HClO_4$ solution, iridium ($i_0 \simeq 10^{-2}$ A m^{-2}) is a better electrocatalyst than Pt, Pt–Rh alloy, and Rh.

The effect of the electrode metal on the exchange current density for the hydrogen evolution reaction in acid media can be seen in Table 4.1. For this reaction, Pt and Pd are the best electrocatalysts.

The electrocatalytic activity of a substrate can also be estimated (see equation (5.12)) from the overpotential required to produce a given

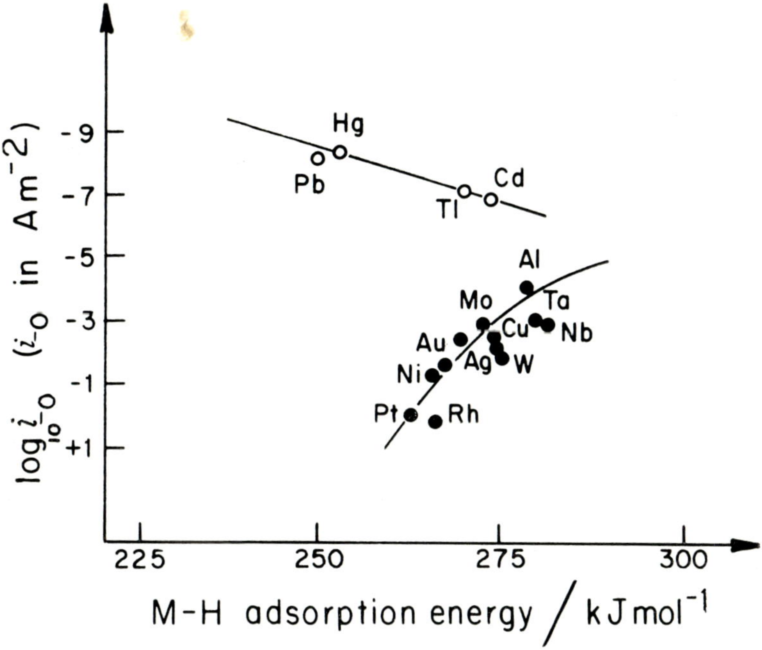

Fig. 5.2. $\text{Log}_{10}\, i_0$ (hydrogen evolution reaction) versus the adsorption energy of hydrogen on the metal.

current density ; the lower the overpotential, the better the corresponding substrate is as electrocatalyst.

Thus, the electrochemical oxidation of methanol in M H_2SO_4 at 373 K at a current density of 200 A m^{-2} requires an overpotential of 0·44 V on platinum black, while on a 70 per cent Pt–15 per cent Ru–15 per cent Mo alloy only 0·23 V are required.

5.3.2. *The genesis of an electrocatalyst*

Electrocatalysts depend on the rate-determining step of the reaction concerned. The direction in which one ought to seek improvement of the catalyst will differ depending upon which rate-determining step is controlling the rate of the reaction. *Any attempt to find improved electrocatalysts for a certain reaction without first finding out what is the rate-determining step in that reaction may be expensively empirical.* If one does know what the rate-determining step is, then one can make some predictions as to which change of type of substrate will increase the rate.

Figure 5.2 shows the relationship between log i_0 for hydrogen as a function of the heat of adsorption of hydrogen on the metal. We see that there are two groups of metals. The first group, the upper one in the figure, is that in which the rate-determining step is :

$$H^+ + e \rightarrow H_{ads}.$$

The higher the heat of bonding to the substrate, therefore, the lower will be the activation energy ; the reaction will tend to go faster because the bond helps the reaction to occur.

For the other group of metals, that represented by most of the transition metals in the diagram, the rate-determining step is an electrochemical *de*sorption according to :

$$H^+ + H_{ads} + e \rightarrow H_2.$$

In this case, it is more difficult for the reaction to occur when the bonding is strong, because it is harder for the hydrogen (i.e. it needs greater overpotential) to desorb from the catalyst surface, as it has to do in forming H_2. Here, therefore, there would be a decrease in the rate of the reaction (at constant overpotential) as the hydrogen bond to the substrate *in*creases in strength.

5.4. *Electrosynthesis*

5.4.1. *The pros and cons of carrying out a reaction electrochemically*

There are several advantages in carrying out a reaction electrochemically.

(i) There are no noxious gases evolved from the reactor. In many cases there are no gases evolved, or the gas is benign, e.g. oxygen, or it can be collected at the electrode. Thus, one of the advantages of electrochemical syntheses is that they contribute little or nothing to the atmosphere.

(ii) Another advantage is selectivity. Electrochemical reactions cannot take place unless there is a significant overpotential. Because the overpotential is given by the equation $\eta = V - V_{rev}$, this means that the reaction begins if the potential is more negative than the thermodynamic potential V_{rev} if the reaction is cathodic, that is, if electrons pass from the electrode to the solution. If it is anodic, with electrons passing from the solution to the metal, the electrode potential has to be more positive than the reversible potential. Theoretically, we can start with one kind of reaction occurring, and then introduce another one by changing the electrode potential. If we want to reduce nitrobenzene to nitrosobenzene, we should first adjust the potential to 0·8 V† (with

† The reversible potential V_{rev} for the reaction $C_6H_5NO_2 + 2H^+ + 2e \rightarrow C_6H_5NO + H_2O$ is $\simeq$0·8 V.

respect to the standard hydrogen electrode) and then make it somewhat more negative, say 0·6 V; but if we wish to produce aniline, we should apply a potential still more negative, say, 0·4 V.

(iii) The activation energies of electrochemical reactions are less than those of the corresponding chemical reactions, so that we should be able to carry out the reactions at a lower temperature, while maintaining the same rate (for *rate* is proportional in chemical and electrochemical kinetics to exp $(-E_a/RT)$, where E_a is the activation energy). This is good, for it avoids the possible difficulties (of corrosion and breakdown of apparatus) involved in operating at a higher temperature.

(iv) Lastly, in chemical syntheses we can only work within a range of energies corresponding to the energies of formation of hydrogen and of oxygen, i.e. we can reduce with hydrogen or oxidize with oxygen. These differ, in electrochemical terms, by 1·2 V, and the energy which corresponds to this is 125 kJ mol^{-1}. This is not very much energy to adjust in trying to make one reaction occur and not another. On the other hand, in electrochemical reactions (utilizing non-aqueous solvents) we can apply positive potentials up to +3 V, and negative potentials down to −3 V. The reason is that in non-aqueous solutions there are not the constituents to give H_2 or O_2 evolution. Thus, we have as much as 6 V available—a greater energy range with which to bring about possible new reactions.

There are two compensating difficulties in electrochemical syntheses. Electrochemical reactions take place *on* an electrode surface, and not *in* a three-dimensional reactor as in a chemical situation. The reaction rate per unit *volume* of an electrochemical reactor is usually much less than that obtained when the same reaction is carried out chemically (although at the same temperature the rate per unit area can be greater). Thus, a unit mass per unit time of an electrochemically produced compound will need more space in technological manufacture. Since land costs money, this disadvantage can be important.

The other difficulty is that the energy has to be paid for. When electrical energy becomes abundant, probably towards the end of the century, this difficulty will be reduced. It is energy which has to be paid for, and that is always a product of *current* × *potential* × *time*. The lower the potential applied to the electrochemical cell from an outside source to make it work, the cheaper will be the product. But this potential is, essentially, E_{rev} (the thermodynamic reversible potential of the cell) together with the overpotentials at the electrodes. The *smaller* these overpotentials are, the cheaper the energy for unit mass of product. *Electrocatalysis determines the economics of electrochemical syntheses.*

We shall give here one example of an electrochemical reaction on a large scale. It is the synthesis of adiponitrile, which is one of the intermediates in the production of nylon.

The basis of the process is the electrochemical reduction of acrylonitrile at the cathode, according to the overall cathodic reaction :

$$2\,CH_2{=}CH{-}CN + 2H^+ + 2\,e \rightarrow NC{-}CH_2{-}CH_2{-}CH_2{-}CH_2{-}CN.$$

The potential at which the reaction occurs at an appreciable rate is more negative than $-1{\cdot}75$ V (on the standard hydrogen scale). To avoid the possible parallel H_2 evolution reaction, mercury or lead cathodes are used (because of their very small i_0 for the hydrogen evolution reaction ; see equation (4.19′)). The yield of adiponitrile is more than 90 per cent, and the main by-product is propionitrile.

CHAPTER 6

electrochemical science in materials science

6.1. *Introduction*

THE properties of materials often depend as much upon what goes on at their surfaces as upon what happens within their bulk. For example, cracks in materials may start at the *surface*, and fatigue failures in materials depend on the precise properties of the *surface*.

Much of Materials Science studies *conditions* for stability and therefore for breakdown ; but the *mechanism* of breakdown and decay of materials, when examined on an atomic scale, is frequently electrochemical.

6.2. *The stability of materials : electrochemical breakdown*

6.2.1. *The basic idea*

Figure 6.1 shows that a corroding metal can be considered as an electron-sink area at which a de-electronation reaction ($M \rightarrow M^{+} + e$) occurs, (2) an electron-source area at which an electronation ($H^{+} + e \rightarrow \frac{1}{2}H_2$ or $\frac{1}{4}O_2 + H^{+} + e \rightarrow \frac{1}{2}H_2O$) occurs, and (3) an ionic conductor that functions as the medium for the electrodic reactions. Such a system is like a short-circuited, self-driving cell.

If the metal on which corrosion is occurring is homogeneous, the electronation and de-electronation reactions occur randomly all over the surface : we speak of the Wagner–Traud mechanism of homogeneous, or uniform, corrosion.

If the metal has particles of other metals on the surface (i.e. impurities), one of the two reactions making up the corrosion cell will tend to take place more at the impurities. This gives rise to *heterogeneous* (or non-uniform) corrosion.

In the first case the sink-to-source distance is less than a nanometre, while in the latter case, the distance between the source and sink of electrons may be of the order of micrometres. In both cases, especially in well conducting electrolyte media, the resistance, and therefore the voltage drop between source and sink areas, can often be neglected. A steady state arises in which the net anodic current, generated by the metal dissolution process $M \rightarrow M^{+} + e$, is exactly balanced by the net cathodic process, say, $H^{+} + e \rightarrow \frac{1}{2}H_2$. As the corroding metal is not attached to an outside source, the net current passing through the metal–electrolyte interface must be zero :

$$I_a + I_c = 0. \tag{6.1}$$

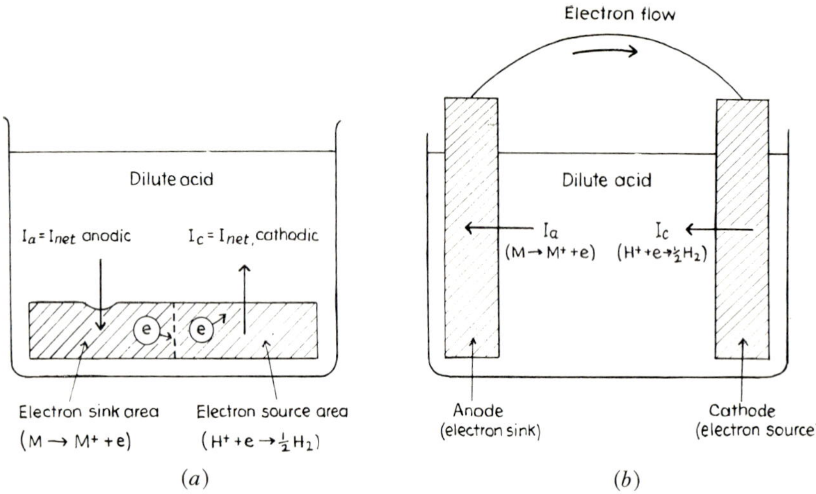

Fig. 6.1. The spontaneous process which forms the basis of electrochemical corrosion. (*a*) The electrons generated by the net anodic reaction at an electron sink area are consumed by the net cathodic reaction at an electron source area. (*b*) A corrosion system is like a short-circuited self-driving cell, having electrodes connected by a wire of negligible resistance. The arrows in the figure indicate the electron flows across the corresponding interfaces.

The steady state value of the anodic current, I_a, is the *corrosion current*, I_{corr}. It expresses the rate of metal wastage by the dissolution process.

6.2.2. *The corrosion potential and corrosion current*

Let us assume that both the cathodic and anodic reactions at the same piece of metal (fig. 6.1 (*a*)) are charge-transfer limited, so that no difficulties arise from transport. Then, the potential–current relationships are given by the Butler–Volmer equation (but in terms of total currents instead of current densities) :

$$I_a = I_{0,1}\{\exp[(1-\beta_1)F\eta_1/RT] - \exp(-\beta_1 F\eta_1/RT)\} \qquad (6.2\,a)$$

$$I_c = I_{0,2}\{\exp[(1-\beta_2)F\eta_2/RT] - \exp(-\beta_2 F\eta_2/RT)\} \qquad (6.2\,b)$$

where the subscript 1 refers to the anodic and the subscript 2 refers to the cathodic reaction. The corresponding overpotentials of the two reactions are :

$$\eta_1 = V - V_{rev,1}$$

$$\eta_2 = V - V_{rev,2}. \qquad (6.3)$$

$V_{\text{rev},1}$ and $V_{\text{rev},2}$ are the equilibrium electrode potentials as given by the Nernst formula (p. 76), and $I_{0,1}$, $I_{0,2}$ the corresponding exchange currents (p. 63).

Let us now apply the steady-state condition (6.1), and for the sake of algebraic simplicity assume that $\beta_1 = \beta_2 = 0{\cdot}5$. Using equation (6.2) with $\eta_1 = V_{\text{corr}} - V_{\text{rev},1}$; $\eta_2 = V_{\text{corr}} - V_{\text{rev},2}$:

$$I_{0,1}\{\exp[\tfrac{1}{2}F(V_{\text{corr}} - V_{\text{rev},1})/RT] - \exp[-\tfrac{1}{2}F(V_{\text{corr}} - V_{\text{rev},1})/RT]\} + I_{0,2}\{\exp[\tfrac{1}{2}F(V_{\text{corr}} - V_{\text{rev},2})/RT] - \exp[-\tfrac{1}{2}F(V_{\text{corr}} - V_{\text{rev},2})/RT]\} = 0 \quad (6.4)$$

After algebraic manipulation,

$$V_{\text{corr}} = \frac{RT}{F} \ln \frac{I_{0,1}\exp(\tfrac{1}{2}FV_{\text{rev},1}/RT) + I_{0,2}\exp(\tfrac{1}{2}FV_{\text{rev},2}/RT)}{I_{0,1}\exp(-\tfrac{1}{2}FV_{\text{rev},1}/RT) + I_{0,2}\exp(-\tfrac{1}{2}FV_{\text{rev},2}/RT)}. \quad (6.5)$$

This expression for the corrosion potential shows that V_{corr} is between $V_{\text{rev},1}$ and $V_{\text{rev},2}$. Consider the case $I_{0,1} \gg I_{0,2}$. Then, equation (6.5) simplifies to:

$$V_{\text{corr}} \simeq \frac{RT}{F} \ln \frac{I_{0,1}\exp(\tfrac{1}{2}[FV_{\text{rev},1}]/RT)}{I_{0,1}\exp(-\tfrac{1}{2}[FV_{\text{rev},1}]/RT)} \simeq V_{\text{rev},1}. \quad (6.6)$$

If equation (6.5) for V_{corr} is introduced into equation (6.2 *a*) for I_{a}, (substituting $V_{\text{corr}} - V_{\text{rev},1}$ for η_1), a complicated expression for the corrosion current results. One simplification which can be made to get the algebra into good shape is to take η_1 and η_2 as both sufficiently large for the Butler–Volmer type equations (6.2 *a*) and (6.2 *b*) to be rewritten in their abbreviated forms (p. 86):

$$I_{\text{a}} = I_{0,1}\exp\{\tfrac{1}{2}[F(V - V_{\text{rev},1})/RT]\}$$

$$I_{\text{c}} = -I_{0,2}\exp\{-\tfrac{1}{2}[F(V - V_{\text{rev},2})/RT]\} \quad (6.7)$$

which leads to:

$$V_{\text{corr}} = \frac{RT}{F} \ln \frac{I_{0,2}\exp(\tfrac{1}{2}[FV_{\text{rev},2}/RT])}{I_{0,1}\exp(-\tfrac{1}{2}[FV_{\text{rev},1}/RT])} \quad (6.8)$$

or:

$$V_{\text{corr}} - V_{\text{rev},1} = \frac{RT}{F} \ln \frac{I_{0,2}}{I_{0,1}} + \frac{V_{\text{rev},2} - V_{\text{rev},1}}{2} \quad (6.9)$$

The expression for I_{corr} is given by equation (6.7) where $V - V_{rev,\,1}$ must be replaced by $V_{corr} - V_{rev,\,1}$. Using equation (6.9) and performing the algebra, one obtains :

$$I_{corr} = [I_{0,\,1} I_{0,\,2}]^{1/2} \exp\left[\frac{F(V_{rev,\,2} - V_{rev,\,1})}{4RT}\right]. \qquad (6.10)$$

Figure 6.2 illustrates the physical meaning of V_{corr} and I_{corr}.

We have assumed :

(1) that only one reaction takes place,
(2) that there is a negligible 'IR' potential drop in the solution,
(3) that $V_{corr} - V_{rev,\,1}$ and $V_{corr} - V_{rev,\,2}$ are large enough and consequently equation (6.7) can be used,
(4) that $\beta_1 = \beta_2 = \frac{1}{2}$, and
(5) that there are no oxide films upon the metal.

However, even from this very over-simplified treatment, some qualitative inferences can be drawn regarding the influence of various factors in determining the corrosion rate.†

6.2.3. *Factors determining the corrosion rate*

Equation (6.10) shows that the corrosion rate depends on the exchange currents and the difference $V_{rev,\,2} - V_{rev,\,1}$.

A metal can be expected to corrode spontaneously only if the difference $(V_{rev,\,2} - V_{rev,\,1})$ between the equilibrium potentials of the anodic metal dissolution reaction and the cathodic reaction is positive. Indeed, only then does the difference $(V_{corr} - V_{rev,\,1})$ become positive, so that a net anodic reaction $M \rightarrow M^{+} + e$ takes place at the electron-sink areas. (In our notation the anodic reaction is given by the subscript '1' and the cathodic reaction by the subscript '2'.)

If $(V_{rev,\,2} - V_{rev,\,1})$ is positive, equation (6.10) indicates that the larger this difference, the higher the corrosion rate will tend to be. For instance, sodium corrodes in aqueous solutions vigorously because the difference $(V_{rev,\,2} - V_{rev,\,1})$ is (for pH = 7) about 2 V.

The exchange current values (I_0) influence the corrosion rate on different metals a great deal (equation (6.10)), because they differ

† Until recently, measurements of corrosion were carried out by immersing the substance in the solution and measuring its loss of mass over months. Now, corrosion measurements are carried out by allowing the metal to attain its steady-state corrosion potential (given by 6.5), and then applying a positive bias potential, so increasing the resulting current. Then a negative bias potential is applied and the resulting current is measured. The logs of the cathodic and anodic currents are plotted respectively against the negative and positive overpotentials, and extrapolated to meet. The log $i_{cathodic}$ (or, of course log i_{anodic}) at the point of intersection is the corrosion current, because this is defined by the current for which both cathodic and anodic currents are equal. Such measurements take a few minutes.

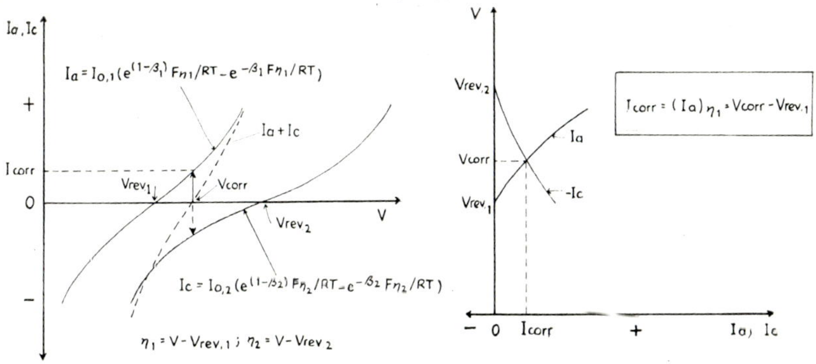

(*a*) (*b*)

Fig. 6.2. The physical meaning of V_{corr} and I_{corr}. (*a*) In the steady state, the net anodic and cathodic currents become equal in magnitude, i.e., $I_a = -I_c$. The net cathodic current is I_{corr}, and the corresponding metal–electrolyte potential difference is V_{corr}. (*b*) A corrosion diagram in $V-I$ coordinates; the coordinates of the intersection point of the curves V vs I_a and V vs $-I_c$, give the values of V_{corr} and I_{corr}.

greatly for various metals. Thus, the exchange current for H_2 evolution on zinc is about 10^8 times greater than on an amalgamated zinc surface (for this has a very low I_0 for H_2 evolution). Consequently, the rate of corrosion of zinc is 10^4 times greater than that of amalgamated zinc.

Equation (6.10) offers a theoretical basis also for the effect of the pH of the solution upon the rate of corrosion of a metal.† The values of c_{H^+} occur in the equation for V_{rev} if that refers to the cathodic hydrogen evolution reaction. $I_{0,\,2}$ (if it refers to hydrogen evolution rather than O_2 reduction) depends on c_{H^+}, too.

The most common cathodic (electronation) reactions in corrosion are hydrogen evolution and oxygen reduction. Which of these two reactions predominates in setting up the corrosion cell is decided as that which gives the largest corrosion current under the circumstances (which depend on the metal, temperature, pH, etc.). One thus inserts numerical values for the exchange currents I_0 and the equilibrium electrode potentials V_{rev} of the anodic and cathodic electrode reactions for both hydrogen evolution and for oxygen reduction, and can thus see, for the condition chosen, which of the two cathodic reactions predominates.

As the standard reversible potential difference V_{rev}^0 (p. 76) is about 1·2 V, while that for hydrogen is zero (and both potentials vary with pH in the same way), it follows that in a given solution $V_{rev,\,2}$ for oxygen reduction is 1·2 V more positive than $V_{rev,\,2}$ for the competing

† pH stands for $-\log_{10} c_{H^+}$; thus, pH$=7$ means $c_{H^+} = 10^{-7}$ mol dm^{-3}.

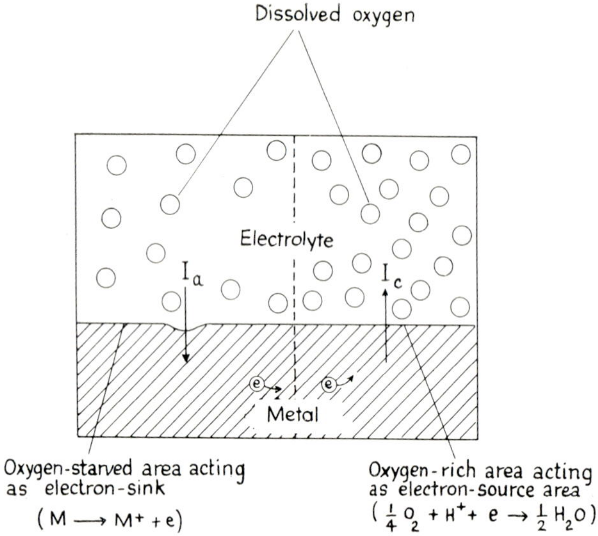

Fig. 6.3. The principle of differential aeration.

hydrogen evolution. The quantity ($V_{rev,\,2} - V_{rev\,1,}$) in equation (6.10) is thus always larger for oxygen reduction as the predominant electronation reaction, which thus results in a much greater corrosion current than the competing hydrogen evolution : oxygen reduction is often the predominant cathodic reaction in corrosion processes. In highly acidic solutions, however, the exchange currents for hydrogen evolution are usually so much larger than those for oxygen reduction that the net effect of both factors $V_{rev,\,2} - V_{rev,\,1}$ and $I_{0,\,2}$ tends to make the hydrogen-evolution reaction the dominant cathodic reaction.

However, as the hydrogen ion concentration c_{H^+} drops, so does I_0 for the hydrogen evolution ; and thus, in concentrated neutral or alkaline solutions, it is the oxygen reduction reaction which usually predominates.

Equation (6.10) also indicates that the corrosion current can be reduced by diminishing the product ($I_{0,\,1} I_{0,\,2}$) ; this is called *corrosion inhibition.*

Thus it turns out experimentally that one can reduce the exchange current, $I_{0,\,2}$, of the electronation reaction by adding, for example, arsenic or antimony compounds to the solution. These compounds adsorb on cathodic sites and reduce the rate of the cathodic reaction (particularly if this is hydrogen evolution) in the corrosion couple. If the electronating or cathodic reaction is oxygen reduction, then a simple way to reduce the corrosion rate is to reduce the concentration of dissolved oxygen. One may simply displace the air-dissolved oxygen from the solution by bubbling nitrogen through it.

A similar approach can be made to reducing the reactivity of the anodic sites. Here, one adds to the solution substances such as amines or thiourea which adsorb on the electron-sink areas and reduce $I_{0,1}$.

Another way to reduce the I_0 values is to reduce the available metal areas, by adding substances to the solution which form a film on the metal surface. For instance, HCO_3^- ions interact with OH^- ions produced by the oxygen reduction reaction in alkaline solution, so that a carbonate film is precipitated on the electrode. This may reduce both $I_{0,1}$ and $I_{0,2}$ although a certain rate of hydrogen evolution may occur *on* the film and metal dissolution may occur slowly through it.

6.2.4. *Typical corrosion situations*

Electrochemical corrosion tends to occur at all metals except the noble ones as long as there is a conducting solution present. But the latter does not have to be a *visible* solution as is present, say, in the corrosion of a ship's hull. It may be an invisible moisture film, with conducting ions in it due to the dissolution of CO_2 or SO_2 from the atmosphere.

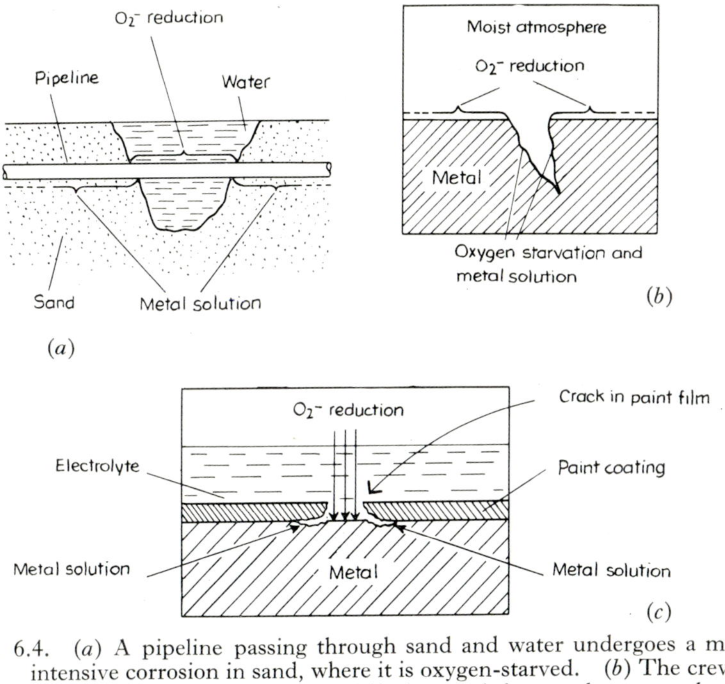

Fig. 6.4. (*a*) A pipeline passing through sand and water undergoes a more intensive corrosion in sand, where it is oxygen-starved. (*b*) The crevice is less aerated, because of slow diffusion, and thus tends to corrode and expand. (*c*) The metal surface underneath the paint coating is the O_2-starved area, and hence it is this protected and invisible part—not the unprotected and visible part—which undergoes corrosion.

One sees at once how atmospheric pollution may reduce the life of metal structures. Electrochemical corrosion is the chief reason why ships have a life of only about 30 years, why bridges need to be constantly repainted (paint protects from moisture and O_2), and why all metal structures gradually decay.

Many examples of corrosion can be interpreted in terms of the *principle of differential aeration* ; if a part of a metal surface has greater access to air (oxygen) than does another oxygen-starved area, then an electrochemical ' corrosion cell ' is set up in which the oxygen-rich area acts as electron-source for the oxygen electronation reaction, and the oxygen-starved area acts as electron-sink for the corrosion reaction so that this is where the metal corrodes. Figure 6.3 illustrates the principle of differential aeration.

Three practical examples of corrosion occurring by differential aeration are given in fig. 6.4.

Corrosion frequently occurs in underground metal structures and pipe-lines because of stray currents arising from a current-carrying cable passing through conducting soils. Because of the electromagnetic field of the cable, a potential difference may be set up along the pipe-line. The regions where the stray induced current in the pipe passes back into the soil become the anodic areas, where metal-dissolution corrosion takes place (see fig. 6.5). At the other areas, the corresponding currents enter the pipe, electrons leave it for O_2 molecules or H^+ ions in the moist surroundings, and such areas of the pipe remain uncorroded. Thus, *patches* of the pipe corrode and eventually break down.

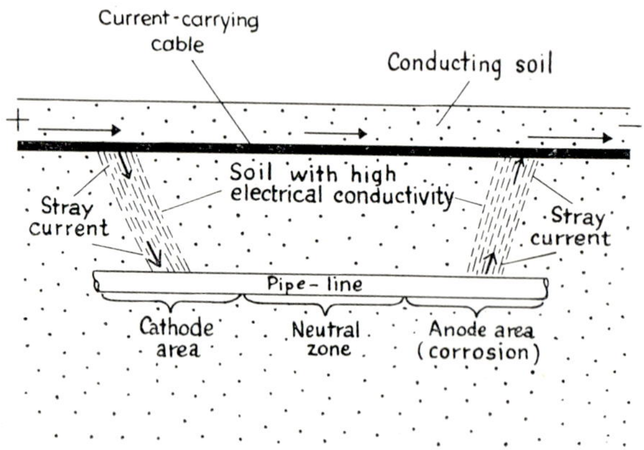

Fig. 6.5. If a pipeline passes through a field of stray currents from a cable, electron-source (cathodic) and electron-sink (anodic) areas may be developed, and the pipeline tends to corrode at the anodic area.

6.2.5. *Cathodic protection*

If the potential difference across the interface at the electron-sink areas could be made negative with respect to the equilibrium potential for the metal-dissolution reaction, no corrosion would occur. This can be achieved by pumping an adequate number of electrons into the corrodible metal, and this method of corrosion prevention is called ' cathodic protection '. It may be effected in two ways :

The *galvanic* method immerses in the corrosive environment an auxiliary metal electrode having a corrosion potential negative to that of the corrodible metal to which it is directly connected. This metal dissolves, and thereby is used up ; it is a ' sacrificial ' anode. It causes cathodic emission from the metal which is to be protected, so keeping this metal protected because its potential is made too cathodic for it to dissolve. Thus, an underwater iron construction can be protected by using sacrificial zinc anodes (fig. 6.6 *a*).

The *generator* method uses an external power supply. The negative pole is connected to the corrodible metal, and the positive pole to an auxiliary inert electrode immersed in the corrosive electrolyte. Thus, the electrons used to make the corrodible metal cathodic (and therefore not corrodible) come from the external power supply. The metal to be protected acts as an electron *source* for the electronation reaction, and the auxiliary inert anode serves as an electron-*sink* area for some de-electronation reaction (see fig. 6.6 *b*).

6.3. *The stability of metals : hydrogen embrittlement*

So far, we have not considered the effect of the adsorption of hydrogen atoms on the surface of the corroding metal which occurs when the cathodic part of the corrosion mechanism is the discharge of protons ($H^+ + e \rightarrow H_{ads}$). They do not simply go off into the atmosphere as hydrogen molecules. In certain metals, some *diffuse into* the metal. This may cause the metal to weaken ; its ductility decreases, it becomes relatively inelastic and prone therefore to break under stress. The metal ' embrittles '.

6.3.1. *The main features of the embrittlement mechanism*

One way in which a metal embrittles is shown in fig. 6.7. Hydrogen deposited upon a metal surface in the course of a corrosion reaction may diffuse inside it. A metal nearly always contains voids, of the order of a few tens of nanometres in length and perhaps several nanometres in the other two dimensions. When H diffuses inside the metal to the walls of these voids, it adsorbs on their surfaces. Some of it desorbs from the surface and passes into the voids ; but, once inside, it is *molecular* hydrogen, H_2.

Within the voids, the hydrogen builds up to a high pressure (10^3 to 10^4 bar), which may exceed the limiting ' spreading pressure ' of the

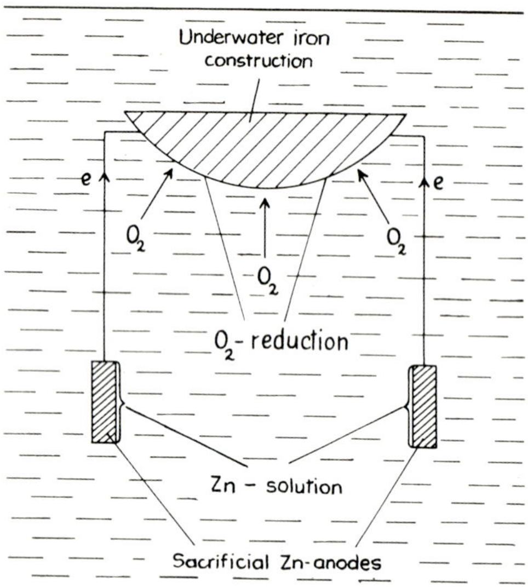

(*a*)

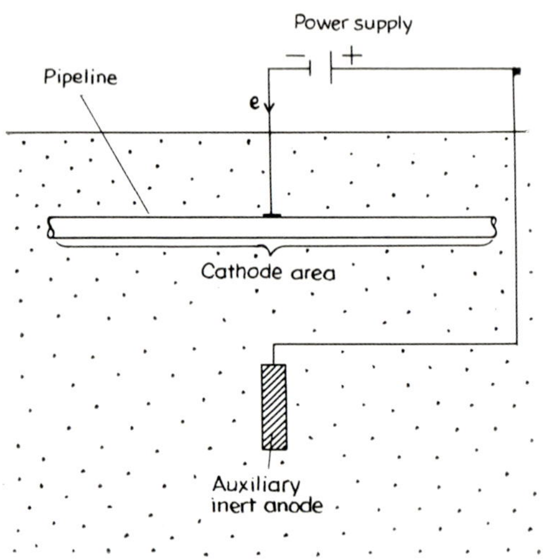

(*b*)

Fig. 6.6. Cathodic protection. Diagram (*a*) illustrates the galvanic method of cathodic protection, using sacrificial anodes. In the alternative generator method, diagram (*b*), inert electrodes are used in place of the sacrificial ones and an external source of power connected between these and the metal to be protected, the latter to the negative pole.

metal constituting the walls of the void. The void then starts to spread.

If such events occur in one void, they will of course occur in many. And if, in the course of spreading, they join up, the metal loses strength.

Why does this high hydrogen pressure develop? Hydrogen evolution on a metal requires a certain *overpotential*. The pressure in the cavities will reach a value *equivalent* to this overpotential. Perhaps we can make an attempt to find out its numerical value by assuming, as a first approximation, thermodynamic equilibrium between the H_2 in the cavity and that on the surface of the metal, where H atoms are being deposited from solution. Now, the difference in the Nernst potential for the hydrogen evolution reaction:

$$2H^+ + 2e \rightleftarrows H_2 \qquad (6.11)$$

at two H_2 pressures, $(p_{H_2})_1$ and $(p_{H_2})_2$ is (equation (5.5)):

$$\Delta V = \frac{RT}{2F} \ln \frac{(p_{H_2})_1}{(p_{H_2})_2}. \qquad (6.12)$$

Suppose the electrode requires a certain overpotential η for deposition from solution to occur. The question is, how can we state the overpotential η in a thermodynamic way, in terms of some change in pressure? The way is given by equation (6.12). We have to set $\eta = \Delta V$ (cf. equation (6.12)) and then take one of the p_{H_2} values as 1 bar (because η is measured with respect to a reference electrode at 1 bar). Thus,

$$\Delta V = \eta = \frac{RT}{2F} \ln \frac{1}{(p_{H_2})_2}. \qquad (6.13)$$

Hence:

$$(p_{H_2})_2 = \exp(-2F\eta/RT). \qquad (6.14)$$

For $T = 300$ K and for an overpotential η of $-0{\cdot}15$ V the excess hydrogen pressure within the void is thus about $10^{4{\cdot}5}$, = about 3×10^4, bar.

There is a certain pressure of H_2 in the voids which will overcome the interatomic forces of the sides of the wall. The void will expand, tear the metal and spread. At what pressure this happens—and, therefore, by equation (6.14), at which overpotential on the outside of the electrode—will depend on the metal, because each material has different interatomic forces to overcome. A closer examination shows that the H_2 pressures inside cracks which cause spreading are indeed in a range given by equation (6.14) with η between about 0·1 V and 0·4 V.

6.4. *The extraction of metals*

6.4.1. *Introduction*

We hear about blast furnaces in steel making, but most metals except iron are extracted electrochemically. Here, the technology is

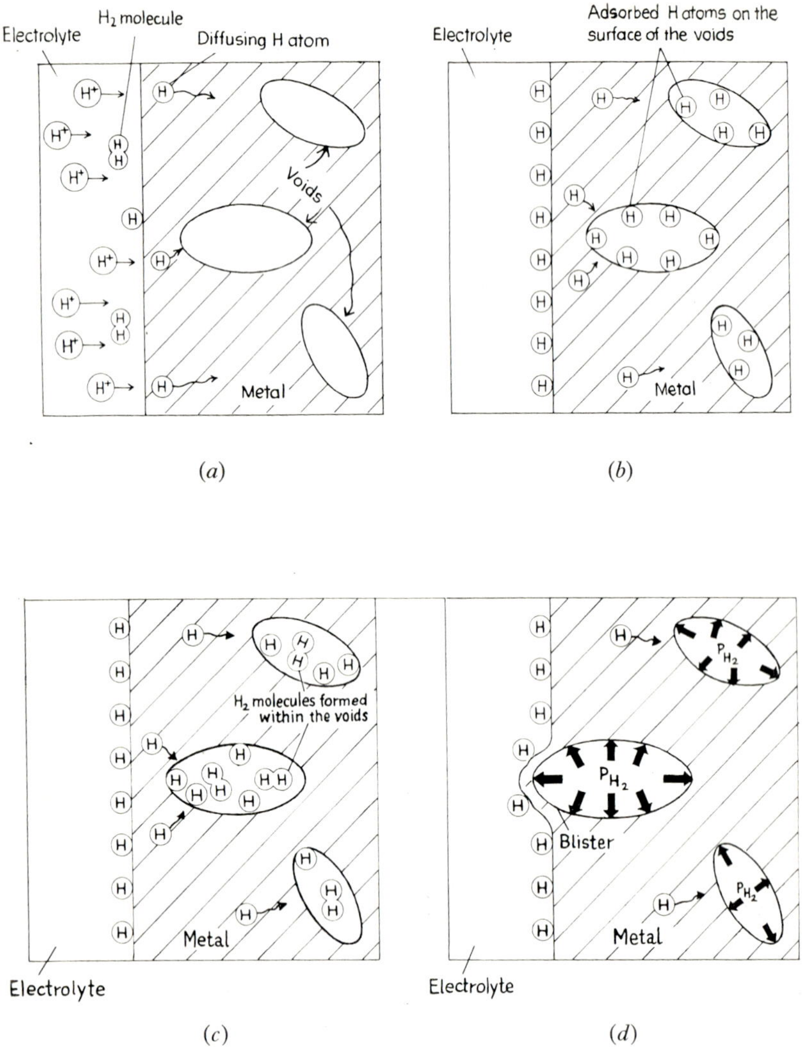

Fig. 6.7. (*a*) Some of the H atoms adsorbed on a metal *diffuse into the metal*, which contains tiny voids. (*b*) After reaching the voids, the H atoms adsorb on the surfaces of the voids. (*c*) H_2 molecules start to form within the voids. (*d*) A high pressure is built up in the voids, the metal near the surface yields and a blister forms. (*e*) Within the solid, the initially small voids spread and meet, giving rise to large cavities connected by fault lines. (*f*) Under stress, the metal no longer behaves in a ductile, elastic fashion, but becomes brittle. It fractures in the surface region. (*g*) Some of the fractures coalesce, forming macroscopic cracks, and causing collapse of a structure under stress.

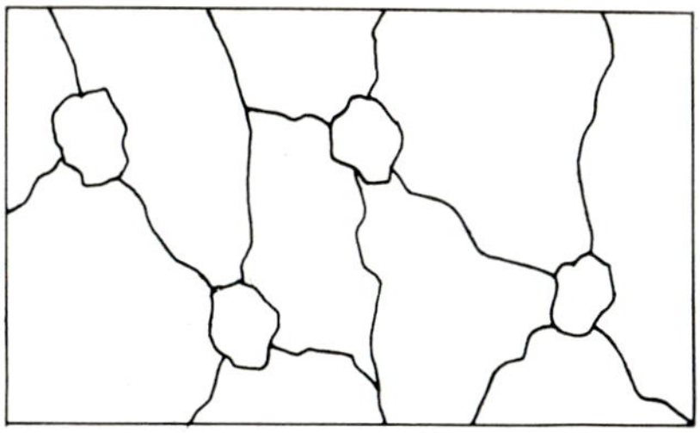

(*e*)

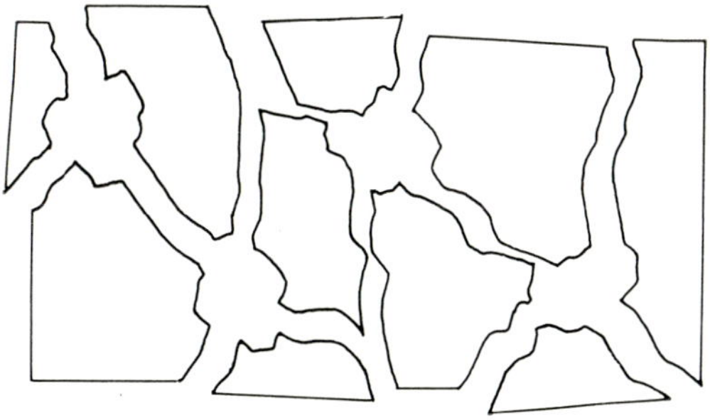

(*f*)

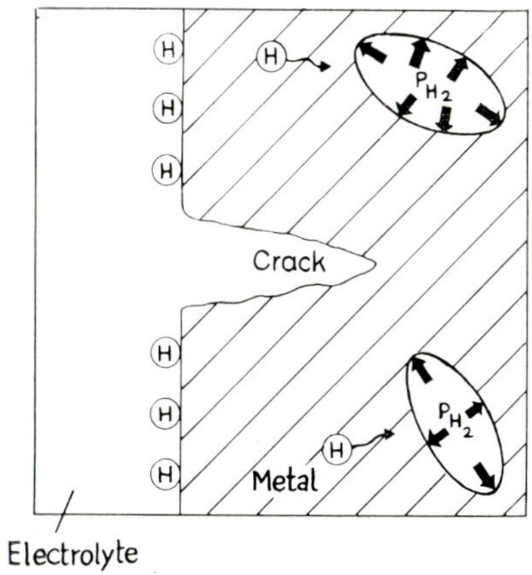

(*g*)

not as new as in other areas of electrochemical science. We shall describe one example only.

6.4.2. *The process of aluminium extraction*

This is carried out from a molten salt called cryolite (Na_3AlF_6); in the molten state at about 1250 K it dissociates into Na^+ and AlF_6^{3-}. About 2 to 8 per cent by mass of alumina, Al_2O_3, is added.

A typical electrolysis cell is shown schematically in fig. 6.8.

When the alumina is added to the melt, complex ions such as $AlOF_2^-$ are found due to the reaction $AlF_6^{3-} + Al_2O_3 \rightleftharpoons 3AlOF_2^-$. This complex ion breaks down and is in equilibrium with Al^{3+}, F^- and O^{2-}. At the interface, Al^{3+} discharges to form aluminium metal. The reaction is simply:

$$Al^{3+} + 3e \rightarrow Al. \qquad (6.15)$$

The anode reactions are much more complicated. During electrolysis, the anodic process consists of the deposition of oxygen, probably from oxygen-donating anions such as $AlOF_2^-$. The positive electrode is carbon and the deposited oxygen reacts chemically with the carbon to form carbon monoxide and then further to carbon dioxide. Thus, the overall anodic reaction may be represented as:

$$3\,O + \tfrac{3}{2}\,C \rightarrow \tfrac{3}{2}\,CO_2. \qquad (6.16)$$

The overall *cell* reaction is:

$$Al_2O_3 + \tfrac{3}{2}\,C \rightarrow 2\,Al + \tfrac{3}{2}\,CO_2. \qquad (6.17)$$

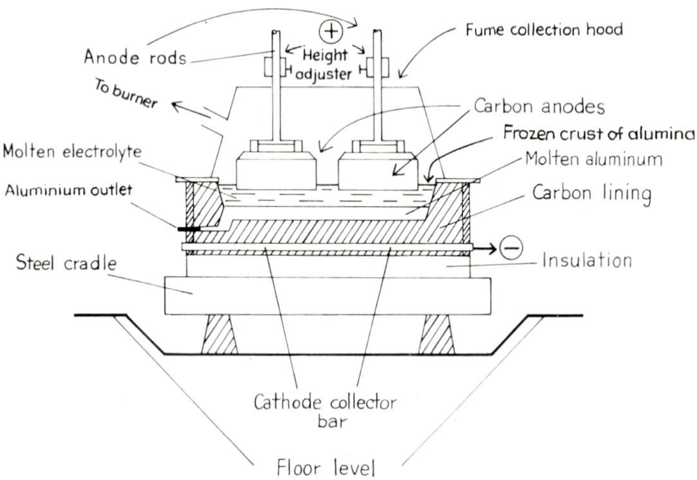

Fig. 6.8. Schematic cross section of an aluminium production cell.

The reversible potential of the aluminium production cell at 1250 K, calculated from thermodynamic data corresponding to this overall cell reaction (i.e. on the basis of the relation $zFV_{\text{rev, cell}} = -\Delta G$, cf. equation (5.11)) is 1·183 V. However, the potential needed to run the cell in practice is much greater than this, because of the overvoltage which is about 0·6 to 0·7 V at a current density of 2000 A m^{-2}. The major portion of this overpotential is due to the anodic reaction. The cathodic reaction requires little overpotential. Then, one has to add the ' *IR* ' potential drop through the molten salt itself which amounts—for a typical 85 000 to 150 000 A commercial aluminium production cell—to about 1·6 to 2·0 V, and also add that across the resistance between the molten aluminium cathode and the cathode bus bar system, which amounts to $\simeq$0·5 V. There are other voltage drops at the contacts in the external conductors, etc., which amount to another 0·5 V. Consequently, the total voltage needed to run the cell is $\simeq$4·7 V, about 3·5 V greater than the reversible thermodynamic potential of the cell. The electric charge that circulates where 1 kg of aluminium is liberated is constant, but the energy required is the product (quantity of electricity $\times$ p.d.), so the cost of the energy needed to produce 1 kg of Al is proportional to the potential (here 4·7 V) which has to be applied to make the cell work at a desired current density.

Thus, application of the results of research on the overpotential of the anodic reaction and above all on the *IR* potential drop in the cell and the potential differences at the contacts, could well lower the price of aluminium significantly. However, the savings indicated by possible new methods or design changes would have to be sufficient to compensate for the cost of the new equipment as well as for the loss of capital due to the replacement of old plant (before it has amortized itself). This kind of economic strait-jacket is not always allowed for by academic consultants to industry, who may think that new ideas are sufficiently justified if they are scientifically correct.

CHAPTER 7

electrochemical science in energy conversion and storage

7.1. *Introduction*

We discussed in Chapter 1 how our energy at present is obtained from the burning of oil, coal and natural gas with oxygen from air. This mode of obtaining energy injects into the atmosphere a constant flow of pollutants, in particular SO_2, NO, CO, and CO_2. Internal combustion engines also emit some unsaturated hydrocarbons, which can cause smog.

Electrochemical energy conversion may be part of the solution to the problem, but the relation between conversion and storage must first be explained. Whether the one or the other, or both, will play an important part during the next 30 years depends on the rate of development of atomic and solar energy sources. If sufficient nuclear energy reactors were to be built by 2000 A.D. there would perhaps be less need for a massive development of fuel cells.

Apart from special situations (see later), fuel cells burning fuels other than hydrogen become important on a massive scale only if governments do not build enough nuclear reactors in time.† In that case, it may be necessary to burn some cheap fuel, perhaps methanol, to produce electrical energy directly by means of electrochemical fuel cells.

7.2. *Fuel cells*

7.2.1. *Fundamentals*

In the hydrogen–oxygen cell—one of the most highly developed fuel cells (p. 10)—the fuel is H_2 and the oxidant O_2: the combustion product is water (fig. 7.1). The electrolyte may be either a strong alkali or a strong acid. A real device is illustrated in fig. 7.2: a concentrated solution of potassium hydroxide is the electrolyte and the catalyst is a heavy-metal oxide, platinum, or a spinel ($Al_2O_3 . CoO$). The electrodes are porous carbon tubes or plates treated with the catalyst.

The electrode reaction at the oxygen electrode (the cathode) in an alkaline electrolyte is :

† The development of solar energy is less advanced than that of atomic energy. There seems little hope that it will be ready before the exhaustion of fossil fuels.

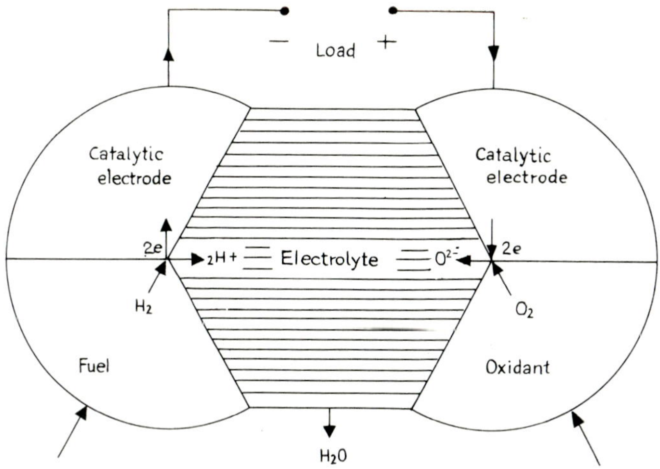

Fig. 7.1. Schematic of a fuel cell : H_2 reacting with O_2.

$$O_2 + 2H_2O + 4e \rightarrow 4OH^-. \tag{7.1}$$

The anode reaction is :

$$2H_2 + 4OH^- \rightarrow 4H_2O + 4e \tag{7.2}$$

so that overall

$$2H_2 + O_2 \rightarrow 2H_2O. \tag{7.3}$$

Many cells are connected in series to attain the required voltages ; each cell gives about one volt.

7.2.2. *The efficiency and the power output of a fuel cell*

By a thermodynamic argument, it turns out that the *ideal* (and hence the maximum) efficiency ϵ_{max} of a fuel cell is :

$$\epsilon_{max} = \frac{\Delta G}{\Delta H}, \tag{7.4}$$

the ratio of the *useful* free energy change, ΔG, of the chemical reaction to ΔH, the change in heat content resulting from the reaction.

For the H_2/O_2 fuel cell $\Delta H = -242$ kJ mol^{-1} and $\Delta G = -229$ kJ mol^{-1}. The ideal efficiency is therefore 229/242, about 94 per cent.

In terms of the reversible equilibrium voltage V_e we can rewrite equation (7.4) :

$$\epsilon_{max} = \left| \frac{zFV_e}{\Delta H} \right| \tag{7.5}$$

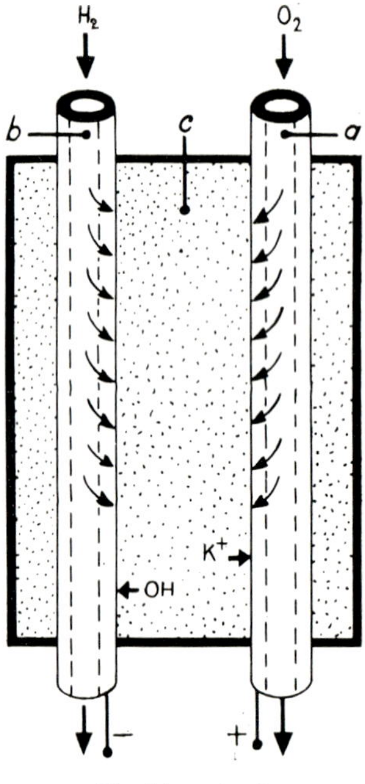

Fig. 7.2. Principle of the H_2/O_2 fuel cell developed by Kordesch *et al.* at Union Carbide Corp. O_2 and H_2 enter through porous carbon tubes (or plates) marked *a* and *b*. The electrolyte *c* is KOH. (*Courtesy of Union Carbide Corp.*, Parma, Ohio, USA).

since from equation (5.11), p. 78, $\Delta G = -zFV_e$, and z is the number of electrons involved in the overall reaction in the fuel cell (for the reaction $H_2 + \frac{1}{2}O_2 \rightarrow H_2O$, $z = 2$).

The *ideal* efficiency referred to above tacitly assumes that no extra energy (over the basic thermodynamic energy) is required to make the reaction go—that is, there is no overpotential. In fact, this is not true: the potential available between the electrodes of the fuel cell when a current is passing is less than the equilibrium (open circuit) voltage, because

(*a*) An appreciable overpotential, at both the cathodic and the anodic reactions, is needed in order to make the electrode reactions (equations (7.1) and (7.2)) proceed at a practicable speed; and

(*b*) The internal resistance of the cell is not zero, so that the passage of the current I through the solution between the electrodes also causes a potential drop, IR_i. Both these effects reduce the useful voltage V_L, which the cell can apply across an external load so that the *real efficiency*, ϵ_{real}, is:

$$\epsilon_{\text{real}} = \left|\frac{zF}{\Delta H}\right| V_L \tag{7.6}$$

where V_L is smaller than V_e by the sum of the voltage drops:

$$V_L = V_e - IR_i - (\eta_{\text{anod.}} + |\eta_{\text{cath.}}|). \tag{7.7}$$

Here, $\eta_{\text{anod.}}$ and $\eta_{\text{cath.}}$ are the anodic and cathodic overpotentials. Both must be taken as positive quantities in equation (7.7), and both depend on the current density between the electrodes, (equations (4.20), (4.20′)). The equations (7·6) and (7.7) should be compared with equation (7.5) which gives the ideal efficiency.

The real efficiency is necessarily below the ideal efficiency (often by about 30 per cent) though it is above that which can be attained with, for example, an internal combustion engine. The real efficiency of the hydrogen–oxygen cell is in the range 58 to 78 per cent, depending on the load current I, while internal combustion engines give about 15 to 25 per cent. The electric power provided by a fuel cell is $V_L I$.

The total overvoltage (cathodic plus anodic) consists of two components at each electrode (p. 73): the charge transfer overvoltage and the overvoltage associated with the diffusion of the ions to the electrodes. Thus,

$$P = V_e I - \frac{IRT}{\alpha_a F} \ln \frac{I}{A_a i_{0,\,a}} - \frac{IRT}{\alpha_c F} \ln \frac{I}{A_c i_{0,\,c}}$$

Max. available power — Power losses due to charge transfer overpotentials

$$- \frac{IRT}{zF} \ln \left(1 - \frac{I}{A_a i_{\text{Lim},\,a}}\right) - \frac{IRT}{zF} \ln \left(1 - \frac{I}{A_c i_{\text{Lim},\,c}}\right)$$

Power losses due to diffusion overpotential

$$- I^2 R$$

Power losses due to the resistance of the solution. (7.8)

where A is the real electrode area, I is the total current iA, α is the transfer coefficient (often $\alpha \simeq \beta \simeq 0{\cdot}5$), i_0 is the exchange current density, i_{lim} is the limiting diffusion current density, R is the gas constant, T the (absolute) temperature, and subscripts c and a refer to the cathode and anode reactions, respectively. From equation (7.8), the greatest power goes with a high exchange current density, i_0, a high limiting current density, a low internal resistance, and a large electrode area.

7.2.3. *Electrocatalysis and fuel cells*

In equations for the energy and power of fuel cells, the exchange current density, i_0, is involved. We have seen that this is connected with electrocatalysis, which is hence a vital factor in developing efficient fuel cells.

Finding good catalysts is a central problem in fuel cell research. Generally, the best catalysts are the noble metals ; platinum is especially good, but is too expensive to use in practical cells. This is an area where much fundamental research is still needed.

7.2.4. *The importance of porous electrodes*

So far, we have shown the importance of increasing i_0 because a large i_0 decreases η, and a low η gives higher efficiency and power. However, mass transport considerations, that is diffusion overvoltages (p. 73), also affect the power available from unit mass of fuel cells. The limiting current density, i_L, is :

$$i_L = \frac{DzF}{\delta} C_{react.} \tag{7.9}$$

and this is the greatest current which can be got from a reaction at a plane electrode. There are only *two* ways of increasing the peak current of the cell :

(1) By raising the *effective* electrode area and/or,

(2) By reducing the thickness of the diffusion layer, δ. This is because $C_{react.}$, the concentration of the electrolyte, is limited by the solubility of the reactant in the solution and because the diffusion coefficients, D, for practical reactants are about the same. If δ is about 0·5 mm, the limiting current density for a 1 M solution is about 400 A m^{-2}. Space is expensive ; an electrode area of 10^{-2} m^2 will yield a maximum current of only 4 A.

One way to greater current densities is to increase mass transport (and hence i_L, see equation (7.9)) by stirring ; this can give an improvement of an order of magnitude, by reducing δ by a factor of about ten. However, part of the increased energy then generated has to be used in working the stirring device.

The most practicable way to raise the current output is the use of *porous*, instead of planar electrodes. The effective electrode area is increased, but the effective value of δ is reduced, so that the limiting current density rises.

In fig. 7.3 (*a*), let the diameter of a pore be $2r$, the surface tension of the (liquid) electrolyte γ, its wetting angle θ_1, and the pressure of the fuel gas (say hydrogen) p. There exists an equilibrium pressure p_e :

$$p_e = 2\gamma(\cos\theta)/r \tag{7.10}$$

at which the meniscus of the liquid within the pore remains stationary.

If the pores are smaller, the capillary effect fills them with electrolyte and blocks the passage of the fuel gas ; the electrode ‘ drowns ’. If

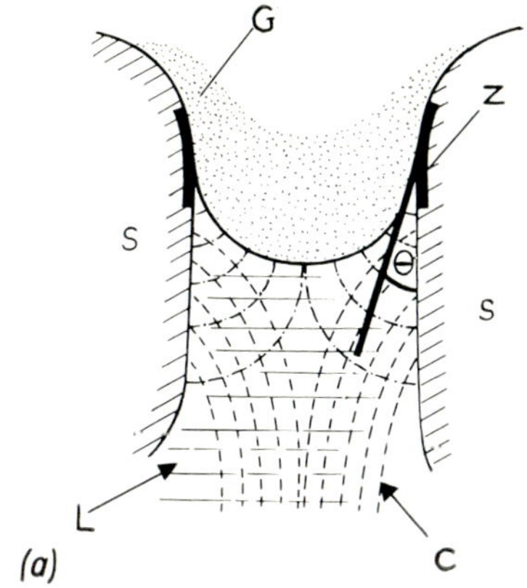

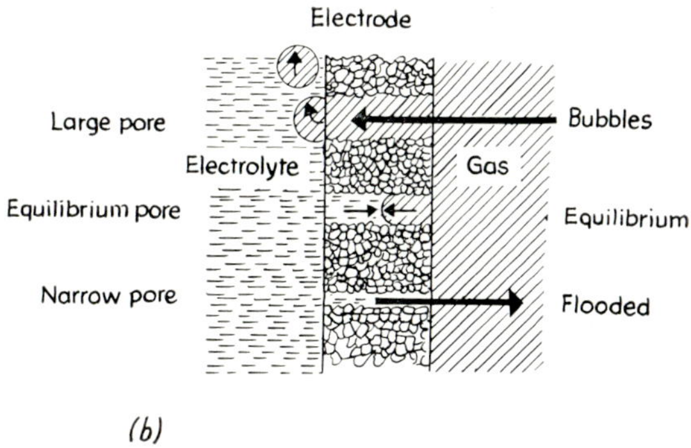

Fig. 7.3. (*a*) Model of a pore. G, gas phase; L, liquid electrolyte; S, solid wall of the pore; Z, active zone; C, diffusion flow lines; θ, contact angle. (*b*) Behaviour of different pore sizes. Schematic cross section through a porous gas electrode showing the different behaviour of large, equilibrium and narrow pores.

the pore is too big, the capillary effect cannot counteract the gas pressure and the gas simply bubbles through an empty pore. This is illustrated in fig. 7.3 (*b*).

In a number of porous electrodes the reaction is confined largely to the meniscus the solution makes in each pore. This is the origin of the increased power from fuel cells which comes with the use of porous electrodes. The meniscus δ is very small, much less than the usual thickness of the Nernst diffusion layer (p. 29). Hence, the effective value of the limiting current is increased (see equation (7.9)), because in place of $\delta = 0{\cdot}5$ mm in equation (7.9) we have the much smaller value, perhaps 10^{-7} m, of the thickness of the reaction zone in the meniscus.

The limiting current in the pore is hence greatly increased compared with the value at planar electrodes.

7.2.5. *The role of fuel cells in the* 1970's

Fuel cells are used in space vehicles, but there are many other possible applications which already make economic sense.

First we might use them in isolated areas, where the cost of energy is practically equal to the cost of transporting fuel. As fuel cells are two to three times more efficient than ordinary sources which burn fossil fuels, we should only have to transport half to one-third of the fuel now needed. The cost of electrical energy would drop correspondingly.

Next, fuel cells could find use in certain vehicles ; for example in submarines, which have to use electrical propulsion when submerged. They run on batteries now, but for submarines in which the time between refuelling is in the region of weeks fuel cells offer a cheaper way to use the fuel. Railway traction, particularly shunting engines, would be another case because fuel cells can easily be turned off and on, and, unlike diesel engines, don't have to waste energy in idling. Likewise, many types of vehicles which work in enclosed spaces, like fork-lift trucks, would gain from fumeless fuel operation. The widest field of use for fuel cells lies in private cars, trucks and buses. There, however, it is not yet clear whether fuel cells or storage batteries would be the best solution. It depends on a race between the exhaustion of the fossil fuels and the provision of new energy sources. But even when solar energy is made available in abundance, it seems likely to be used to electrolyse water and give H_2, which would then be stored and shipped (in pipes as gas ; or in tankers as liquid) to the places where it is to be used. There it would give electrical energy from fuel cells which seem, therefore, to be the likely motive power for passenger transport.

7.3. *Primary and secondary batteries*

7.3.1. *Introduction*

We call rechargeable electricity storage devices *secondary batteries* or *accumulators*. Electrochemical generators of electrical energy which can't be recharged we shall call *primary* batteries. Primary batteries are really a bit like fuel cells, because they convert chemicals into new materials, and yield electrical energy on the way.

Secondary batteries have a different goal from that of fuel cells. The latter are *converters*, they produce electrical energy which didn't exist before from the energy given out when chemicals react. The rechargeable battery should be regarded as a storer, not a converter. The electrical energy has to come from somewhere else—a solar energy platform, say—and the secondary battery accepts it, chemical changes occur on its electrodes, and then, when the switch is turned on, the chemicals try to revert to their original state and give back the electrical energy.

7.3.2. *What makes a 'good' battery?*

The performance of a battery may be judged mainly by (*a*) energy density, (*b*) power density. The energy density is the amount of energy which can be stored per unit mass. Thus:

$$\textit{energy density} \text{ in J kg}^{-1} = (\textit{charge storage density}) \times V_L \quad (7.11)$$

where V_L is the effective voltage across the cell (p. 103). The ideally attainable maximum energy density is a well defined quantity, but the *practically* realizable energy density depends on several electrode kinetic considerations such as the availability of easy current paths in the interstices of crystalline electrode materials, so that a high proportion of the crystal mass is *active* in the charge and discharge processes. The practically realizable energy density of a battery decreases as the rate of using up of the battery increases. It is often less than one-quarter of the ideal maximum energy density.

The power density is the *rate* at which energy can be supplied from the battery per unit mass. But power is the product of the maximum current I_{max} and V_L: thus

$$\textit{power density} = I_{max} \cdot V_L/\text{mass}. \quad (7.12)$$

7.3.3. *Primary batteries*

The primary cell is really a fuel cell where the fuel is held in or on the electrodes instead of in a tank. The electrodes therefore are being consumed in the discharge process, and the life time of the cell is limited. Primary cells cannot be recharged; they are storers of electrical energy which comes from elsewhere. However, one cannot feed fuel to them from a tank, as with fuel cells, so they are not *continuous* energy converters either.

The battery used for electric torches is a primary battery. The cell consists of a zinc anode and a manganese dioxide cathode in contact with an inert collecting electrode of carbon. The electrolyte is a solution of ammonium chloride and $ZnCl_2$ usually in the form of a gel. It is a device of low energy density, about 30 000 to 100 000 J kg^{-1} depending on the discharge rate.

The electrochemistry of these cells is complex, but for low current densities the reactions are probably:

$$\begin{array}{ll} \text{Anode} & Zn \rightarrow Zn^{++} + 2e \\ \text{Cathode} & 2MnO_2 + Zn^{++} + 2e \rightarrow Zn\,Mn_2O_4 \\ \hline \text{Overall} & Zn + 2MnO_2 \rightarrow Zn\,Mn_2O_4 \end{array} \quad (7.13)$$

The terminal voltage at room temperature is about 1·5 V, if the discharge rate is low.

7.3.4. *Secondary batteries* (*accumulators*)

A fuel cell takes some fuel—for example methanol—and *converts* it with the aid of some other material—for example oxygen—to produce electrical energy. It does so for as long as fuel continues to be supplied.

A secondary battery (accumulator) *stores energy* in the form of chemical energy, which it then reconverts into electrical energy upon demand. It accepts energy in the *charging cycle* which *forces* an electrochemical change within the cell. The battery can then be *discharged*; the electrochemical changes are reversed and now occur spontaneously. The substances formed during the charge cycle cause electrochemical reactions resulting in (i) the release of electrical energy, (ii) a return to the initial state of affairs. The most familiar secondary battery is the lead–acid one used in motor cars.

The electrodes are porous lead and porous lead dioxide in an electrolyte of fairly concentrated sulphuric acid. The electrode reactions in the discharge cycle are:

$$PbO_2 + 4H^+ + 2e \rightarrow Pb^{2+} + 2H_2O. \tag{7.14}$$

The formation of Pb^{2+} in the electrolyte causes insoluble $PbSO_4$ to form within the porous material.

The main advantage of the lead–acid battery is that it can deliver a heavy current for a short time, which makes it an excellent choice for the starting motors of cars. This high power density arises because the exchange current densities of both anodic and cathodic reactions are quite large. As an energy storage device, however, its performance is poor, its energy density being about 54 000 J kg^{-1}, whereas the energy density of many batteries developed in the 1960's is five to ten times greater than this (table 7.1).

System	Overall electrode reaction	$V_{rev.}$ at 298 K	Cell voltage at load current 1000 A m^{-2}	Single cell power density W kg^{-1}	Single cell energy density kJ kg^{-1}
Pb–acid		2·10	1·7	110	72
Li–Cl	$2Li + Cl_2 \rightarrow 2LiCl$	3·53	2·9	275	1260
Na–S	$Na + S \rightarrow Na_2S$	2·08	1·9	250	1050
$Zn–O_2$	$2Zn + O_2 \rightarrow 2ZnO$	1·62	1·1	40	525

The values vary with the rate of discharge. The value approximately halfway through the range is given here.

Table 7.1. Performance characteristics of some modern batteries.

Among the best ionically conducting materials are *molten salts* : the high temperatures associated with them (>600 K) mean that the exchange current densities are large (because all rate constants increase with increase of temperature) and thus, cf. equation (4.18), the overpotentials are low.

Several batteries are being developed which use such media as the electrolyte. They have the added advantage that, not involving water, they can use highly reactive alkali metals like Li to produce high cell voltages as well as high energy densities because the free energy change (p. 78) in cell reactions involving alkali metals is high. Thus, the Li–Cl system has an electromotive force of about 3·5 V, its terminal p.d. on open circuit. The reactions of this cell are :

$$\begin{array}{ll} \text{Anode} & 2Li \rightarrow 2Li^{+} + 2e \\ \text{Cathode} & Cl_2 + 2e \rightarrow 2Cl^{-} \\ \hline \text{Overall} & 2Li\ (\text{liquid}) + Cl_2\ (\text{gas}) \rightarrow 2LiCl\ (\text{fused salt } 900\ \text{K}) \end{array}$$

Systems such as these do require an initial melting by some auxiliary source, such as a small lead–acid battery, otherwise the solid electrolyte remains frozen.

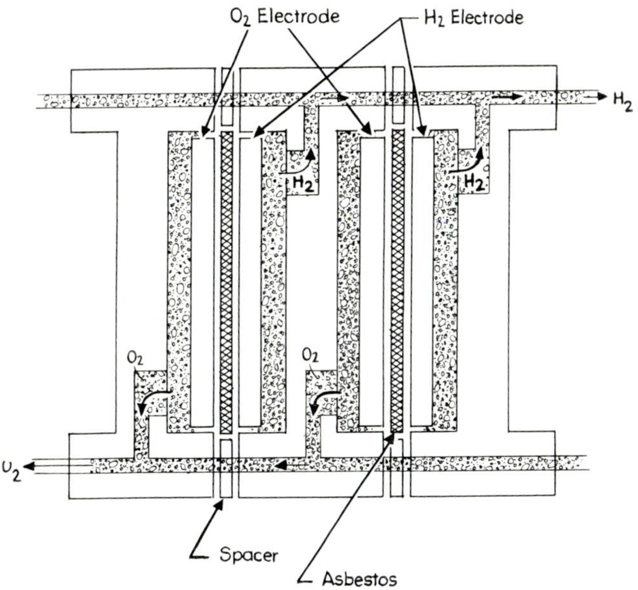

Fig. 7.4. Schematic of a regenerative H_2/O_2 fuel cell. (After E. Findl and M. Klein.)

7.3.5. *The new batteries*

During the last decade, mainly stimulated by the U.S. space programme, many new batteries have been developed at a research stage. There is a tremendous difference between the energy density of the lead–acid battery and those of the newer batteries, as can be seen from table 7.1. The newer batteries offer really striking performances.

If we combine a H_2/O_2 fuel cell with a water electrolysis cell, we obtain a regenerative H_2/O_2 system (fig. 7.4). In charging, water is electrolysed to obtain hydrogen and oxygen ; the H_2 is stored. Upon discharge, the H_2 is fed into a H_2/O_2 fuel cell with O_2 from the air to produce electrical energy, forming water as the reaction product, which is stored for later electrolysis. The same Ni electrodes serve in both the charging and discharging cycle.

Other promising new battery systems include the Li–Cl, Na–S, and the Zn–air cells ; their performance data are also given in table 7.1. These batteries have been explored only in the last few years. They are not yet commercially viable, partly because they require much more research, e.g. on the associated corrosion problems.

A world-wide attack upon the development of high energy electricity storers should claim high priority in the research effort of developed countries.†

If we could develop a satisfactory energy storer which would give us, say, at least 700 kJ kg^{-1} and 200 W kg^{-1}, we could then use solar-derived electricity to run the transportation system without any pollution at all.

† In Japan, this was set up in 1971 as one of twelve national goals for research.

CHAPTER 8

electrochemical science in ecology

8.1. *Introduction*

TECHNOLOGICAL man oxidizes, and then throws out as waste, the degraded resources of the earth into the sky and the sea. It is now realized that this cannot be continued. The dilemma is that the science which man has used so blithely for 100 years has made Nature so much less hostile that there has been a vast increase in population. There is no way back to a life independent of machines, even if it were desired by a majority, without provoking disaster by starvation. Conversely, man cannot continue the present technology, or he will provoke a disaster of another kind : the poisoning of the planet and its population by air and water pollution, uncontrolled breakdown from the exhaustion of resources, or above all, lack of fuels to energize the technology. One or the other end threatens within one to two generations from now.

Man's only hope is to convert his present fossil fuel-based technology to one run entirely by electrical energy, originally derived from solar, nuclear or (as a remote possibility) geothermal sources. If he does this in time†, he can survive, and may even be able to attain a high material standard of life for the whole world population. The only absolutely unavoidable requirement is that we must stop population growth.

8.2. *The burning of fossil fuels and the carbon dioxide balance*

The carbon dioxide balance in the atmosphere has been kept roughly constant over many millions of years. The carbon dioxide ejected into the atmosphere as a result of metabolic processes, and other natural processes, has been converted into oxygen again by photosynthesis, the process by which the energy of sunlight is used by green plants in the conversion of CO_2 into carbohydrates. The reaction can be written :

$$CO_2 + H_2O + h\nu \rightarrow (CH_2O) + O_2 \qquad (8.1)$$

where (CH_2O) is a generalized formula for carbohydrates, $h\nu$ is the energy of a photon of frequency ν and h is the Planck constant.

However, at the beginning of this century, man began to upset the balance appreciably by ejecting into the atmosphere increasing amounts

† It is not a matter of whether he can, technologically, (for he can), but rather whether the realization of the necessity for great change will be accepted by society in time.

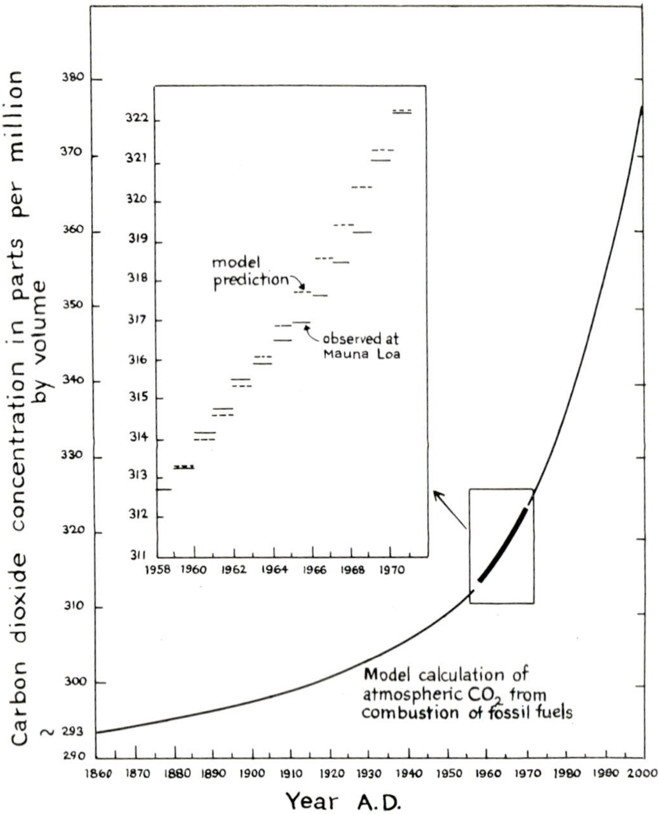

Fig. 8.1. Carbon dioxide concentrations in the atmosphere. Atmospheric concentration of CO_2, observed since 1958 at Mauna Loa, Hawaii, has increased steadily. Source: D. H. Meadows *et al.*, *The Limits of Growth*, Universe Books, NY 1972.

Change in CO_2 content, part per million	Resulting temperature change/K
from 150 to 300	1·3 to 2·8
from 300 to 600	1·33 to 2·92

Table 8.1. The calculated changes in the equilibrium temperature of the Earth due to change in the CO_2 content of the atmosphere.

of carbon dioxide from his unfortunate method of obtaining energy via the heat from burning coal and oil. Figure 8.1 shows that the green plants have been unable to cope with the excess CO_2 since about 1900.

The earth's heat balance is affected by the fact that much of the sun's radiation striking the earth's surface is of a wavelength range that can penetrate the atmosphere fairly well, but after it has been absorbed, and heated the earth's surface, the radiation outwards from the earth is mainly in the long infrared, in which carbon dioxide has strong absorption bands. So the atmosphere is heated by the outward-going radiation, to an extent which depends on the proportion of CO_2 it contains (see table 8.1 and fig. 8.1). The values were calculated under various assumptions as to the condition of the sky, whether clear or cloudy, and as to humidity.

This disturbance of the planet's radiation balance should cause melting of some of the ice in the Antarctic and Greenland. The mass of water thus released, added to the oceans, could cause their level to rise by about 30 cm by about 2000 A.D., according to some computations. After that it could rise at an increasing rate, if CO_2 continues to be injected into the atmosphere at rates extrapolated from the present curve of the CO_2 concentration versus time.

The effect of pollution by *particles*—dust and smoke from fossil fuels, asbestos from brake linings, rubber particles from tyres and so on—introduces uncertainty into these calculations, since it would reduce the solar radiation reaching the earth. It is unlikely that it would just balance the temperature rise caused by the 'greenhouse' effect, but it should retard it. At present, we don't know which effect (heating or cooling) will predominate by 2000 A.D., but the point is that fossil fuel burning *will* disturb the climate significantly by that time.

8.3. *Fossil fuel burning and the oxygen balance*

Most of our oxygen is being continuously replenished by green plants. This process, too, is being disturbed. Forests are being replaced by cultivated fields, reducing the area of leaf surface for photosynthesis. Pollution of the oceans may interfere with the photosynthesizing microscopic sea-life (e.g. various forms of unicellular algae) which are a more important source of oxygen than are the green plants on land.

The pollution of ocean waters is serious now: the land-locked Baltic Sea is close to being a 'dead sea' as is seen from the oxygen depletion in its waters (fig. 8.2).

The conclusion from the last two sections is plain: man must stop burning fossil fuels and look entirely to atomic or solar sources for his energy supplies. If the attempts of automobile manufacturers and oil producers to 'clean up the internal combustion engine' succeed, this will merely *reduce* the pollutive output of certain substances; these will still continue to build up in the atmosphere, though at first at a slower rate. Carbon dioxide, which is the major constituent of the exhaust gases, will continue to rise in concentration for as long as the combustion of fossil fuels exceeds the regenerative capacity of

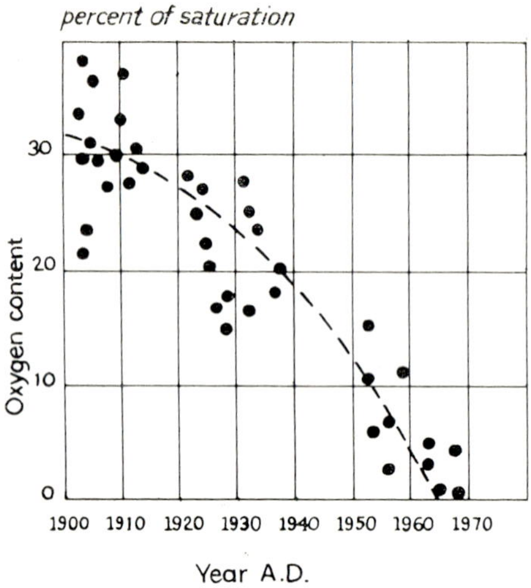

Fig. 8.2. Oxygen content of the Baltic Sea. Increasing accumulation of organic wastes in the Baltic Sea, where water circulation is minimal, has resulted in a steadily decreasing oxygen concentration in the water. (Stig. H. Fonselius, *Environment*, July/August 1970.)

natural photosynthesis. The unsaturated hydrocarbons produced will continue to react with the NO_2 also produced to form 'smog'—the condensate on particles of dust and soot of the peroxyacylnitrates which originate in the hydrocarbons of car exhausts.

What has all this got to do with electrochemistry? The connection is direct. It seems likely that we shall have to carry the energy to drive our cars on board them in batteries. However, should government policies prevent the timely development of alternative energy sources required for charging batteries on a massive scale, we may have to obtain our motive power via fuel cells for a few decades. Fuel cells are twice as efficient as internal combustion engines in converting

Metal	Known global reserves/10^6 tonne	Prospective lifetime /year
Copper	308	36
Lead	91	13
Silver	28	16
Tin	4·3	17
Tungsten	1·3	40
Zinc	120	23

Source: U.S. Bureau of Mines, *Mineral Facts and Problems* (Washington D.C.: Government Printing Office, 1970).

Table 8.2. Some non-renewable natural resources (metals).

chemical energy to mechanical energy, so we could eke out our remaining supplies of fossil fuels (burning half as much for the same distance travelled). Pollution would be much less, too: no unsaturates and smog, although SO_2 and CO_2 would still be produced (e.g. from fuel cells run on coal), but only at half the present rate per car-kilometre.

8.4. *Recycling of metals*

Table 8.2 lists the approximate extent and projected lifetime of some of our non-renewable resources. If we continue to run our technology by taking up raw materials from the earth without recycling them, we shall exhaust our reserves of most metals by the end of the century. For some metals (Pb, Ag) we are within years, not decades, of that happening. The crisis will be felt in advance of the exhaustion: as the richer, more easily worked deposits vanish, we shall have to mine second grade ores at a cost which will bring the problem home to all citizens.

We must treat materials as Nature has been treating water since it was first formed on this planet. It is used, rejected into rivers, lakes and oceans, whence it re-evaporates forming clouds which rain down again as fresh water. Metals are not volatile, so that we shall have to recover them, not by any distillation process (quite impracticable because of the high temperatures needed, for a start) but by some other separation method. Many of these other methods seem likely to involve electrochemical treatment of the metals in solution.

The principle of separation by electrolysis is straightforward; metals begin to deposit at different potentials according to the *electrochemical series*. If the potential of the working electrode is more negative than the standard potential given in fig. 8.3—but *not until then*—the metal begins to be deposited. For deposition at a practical rate, at a definite current, not only the reversible potential must be attained, but also some overpotential has to be added (equation (4.18), p. 64). The key instrument is a potentiostat. This is an electronic control device which sees to it that the electrode potential is kept constant. Thus, the deposition of one chosen metal is assured.

One of the problems of our motorized civilization is what to do with scrapped cars. All the metal content of Table 8.3 could be recovered, and even profitably, since the scrap metal value of a car is in the region of £70. One would first remove the paint by burning; the rest of the car could then be treated by dissolving the compacted (or shredded) car body in an electrolysis bath and then depositing the constituents selectively.

Similar methods could be applied to other used metal articles at present discarded. Of course, in some instances, a different approach may prove advantageous. What is certain is that all our metals *will have to be recycled.*

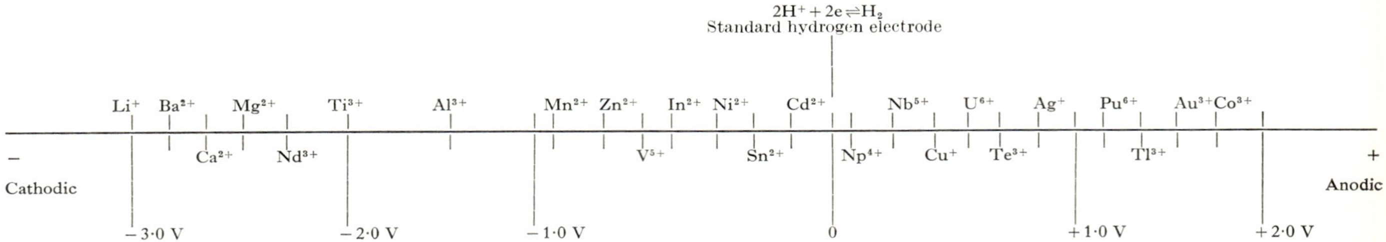

Fig. 8.3. The Electrochemical Series. The potential of the standard hydrogen electrode defines the arbitrary potential zero.

Metal	Mass/kg	Price/£ kg^{-1}	Value/£
Iron	1350	0·04 ‡	54·00
Copper	14·4	0·44	6·40
Zinc	24	0·12	2·90
Aluminium†	23	(0·64)	(14·70)
Lead	9	0·12	1·10
Nickel	2·3	1·00	2·30
Chromium	2·3	2·40	5·50
		Total	£72·20

† Not recoverable in aqueous solution. Price is quoted for scrap, but excluded from total.
‡ Price for top grade scrap.

Table 8.3. The major metals in a motor car.

8.5. *Purification of liquid factory waste*

Factory effluents cannot be rejected into rivers, lakes and oceans. They must be purified first, and that can also be done effectively by electrochemical means.

As an example, consider the disposal of cyanide wastes.

Cyanide compounds find wide application in metal finishing, in the heat treatment of metals, and in chemical industry. Electrochemical, practical processes exist for the conversion of the poisonous cyanide wastes to compounds safe enough to be rejected into a public waste system.

The electro-oxidation proceeds as :

$$2CN^- \rightarrow 2CN + 2e \qquad (8.2)$$

$$2CN \rightarrow (CN)_2 \qquad (8.3)$$

$$(CN)_2 + 2NaOH \rightarrow NaCN + NaCNO + H_2O. \qquad (8.4)$$

Further reactions with added alkali yield non-toxic end products $(NH_4)HCO_3$, Na_2CO_3 and $NH_{3(aq)}$.

Analogous processes could be developed for other factory effluents.

The above treatment employs the principle of anodic oxidation to convert objectionable polluting liquids into something harmless. However, we can use another method, namely electrodialysis.

Some membranes contain fine pores which exhibit diffuse double layers on their surfaces which carry excess negative charges (p. 105) in the solution. Other types of membrane carry positive charges. Application of an electric potential difference across the membrane causes the first type to pass anions preferentially while the other passes cations preferentially. In a cell such as that in fig. 8.4, application of an electric field causes a separation because both anions and cations move outwards away from the central solution, which becomes denuded of its salt-like materials. Thus, one could avoid the need to bring fresh water to factories, and the nuisance of rejecting effluent into rivers. Water can be circulated continuously, being continuously repurified by electrodialysis.

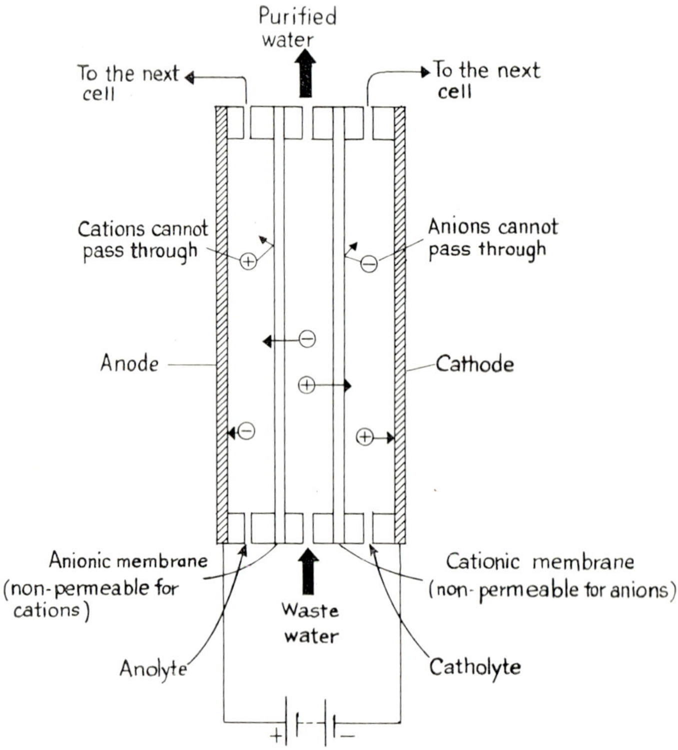

Fig. 8.4. An electrodialysis cell.

8.6. *Clean metallurgical processes*

Those who have seen copper smelters in action will have noticed the enormous clouds of sulphur dioxide vented into the atmosphere. In the present chemical extraction of copper, the overall reaction for copper pyrites (mainly $CuFeS_2$) is :

$$2CuFeS_2 + 5O_2 \rightarrow 2Cu + 2FeO + 4SO_2. \quad (8.5)$$

Such pollution of the atmosphere is unnecessary. Metal sulphides can be treated electrochemically in a way which avoids such air pollution. In this process, the ores are electrolysed in a brine solution, after prior crushing and grinding. Chalcocite is thus anodically oxidized to cupric salts and sulphate :

$$Cu_2S + 5OCl^- + 2H^+ \rightarrow 2Cu^{2+} + 5Cl^- + SO_4^{2-} + H_2O, \qquad (8.6)$$

where the hypochlorite ion is formed from the brine :

$$2Cl^- - 2e \rightarrow Cl_2 \qquad (8.7)$$

$$Cl_2 + H_2O \rightarrow OCl^- + 2H^+ + Cl^-. \qquad (8.8)$$

Excess Cl^- in the bath yields chloro-complexes of copper such as :

$$CuCl_2 + 2Cl^- \rightleftarrows CuCl_4^{2-}. \qquad (8.9)$$

The metal is then recovered by precipitation with zinc.

$$Cu^{2+} + Zn \rightarrow Cu + Zn^{2+}. \qquad (8.10)$$

The method is applicable to other sulphide ores, such as those of silver, molybdenum, antimony, and particularly mercury.

8.7. *Abundant electrical energy*

The promise of a clean electrochemical future depends upon the availability of a supply of abundant electrical energy. The best way to obtain this energy would be by means of solar energy, because that way would be completely non-polluting. However, so little research is being funded in the world that there is little hope that such systems would be available by 1990. Until governments realize the need for solar energy development, and fund it in effective amounts, we shall have to construct nuclear power stations. The cost of the plant required to generate one unit of energy electrically drops with increasing size. By constructing sufficiently large reactors, according to an estimate by Hammond, the cost of electrical energy in the U.S.A. could be reduced to about one-fifth of present cost by the year 2000A.D. However, there is no government policy to build such reactors in sufficient quantity at this time.

Much radioactive pollution arises from nuclear reactors, and this is a potent reason *against* the construction of nuclear reactors on a vast scale, as seems (in 1973) to be the way things are tending. One method of construction would involve incorporating them in heavy concrete containers or blocks which could then be put into deep disused mine shafts, although these containers could still corrode and leak.

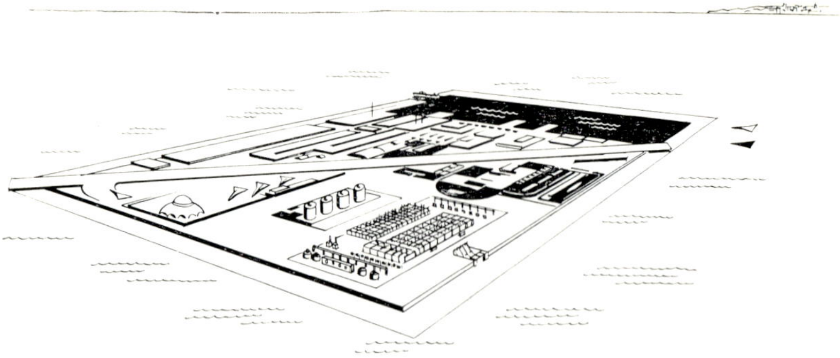

Fig. 8.5. A floating energy centre.

Reactor stations will have to operate under the laws of thermodynamics, so they will still produce thermal pollution, as they have to reject their waste heat. This would be unacceptable on land, except perhaps in central Australia, Northern Canada, etc. It could be readily dissipated at sea : a speculative impression of an energy centre is shown in fig. 8.5. It would float some distance off shore. The water depth should be sufficient for deep water, pumped up for cooling, to be sufficiently cold. The cost of electrical energy actually *on* such a floating Energy Island would be below that on shore, and therefore industry would be attracted to build plants there, the more so because there would be no limitations on the draught of vessels bringing supplies in bulk. The same ideas would apply if the islands were solar energy platforms.

8.8. *The hydrogen economy*

About 50 per cent of the consumer price of electrical energy is transmission cost. Current has to be sent through wire cables of finite resistance, and the resulting voltage drop represents energy dissipated as heat. It would be cheaper to generate hydrogen from water at the reactor, send the gas through pipes, and regenerate electrical energy by *fuel cells* at the site of use. Where heat is needed, the hydrogen could be burned to water.

The most important advantage of this means of supplying us with energy would be an avoidance of pollution. However, there would be another advantage. The hydrogen–oxygen fuel cell yields drinkable water as a by-product. By about 2000 A.D. the amount of electrical energy expected to be used by the average household would produce (if supplied by H_2–O_2 fuel cells) enough fresh water for the drinking and cooking needs of that household. A *hydrogen economy* (energy centres →electricity →hydrogen →pipe →fuel cell conversion at site) would give abundant and clean energy, fresh water, and very cheap excess hydrogen for use in a cleaner chemical industry.

The hydrogen could be used to run the transport system electrochemically, with fuel cells driving the motors, and provide clean alternatives to present metallurgical processes. For example, iron ore could be reduced directly :

$$Fe_2O_3 + 3H_2 \rightarrow 2Fe + 3H_2O, \tag{8.11}$$

with no CO or CO_2 rejected from blast furnaces, as at present.

CHAPTER 9

electrochemistry in biology

9.1. *Introduction*

Biological systems are highly organized in that they involve many structural entities, of which nearly every one is contained by a *membrane* which is defined by two interfaces : one toward the inside and the other toward the outside. Electric double layers and electric potential differences (Chapter 3) arise wherever there are interfaces.

9.2. *Membranes*

A simplified sketch of a biological membrane is shown in fig. 9.1. The main structure consists of a ' bilayer ' of lipo-protein molecules oriented so that their hydrophilic heads point outwards while their hydrophobic tails point inwards. Polar groups on the membrane surfaces cause water molecules (not shown) to be attached in highly oriented layers. Pores—only one is shown—penetrate the membrane ; they are filled with polarized water molecules and their insides carry charged groups. Counter-ions, such as K^+, are attached to the polar

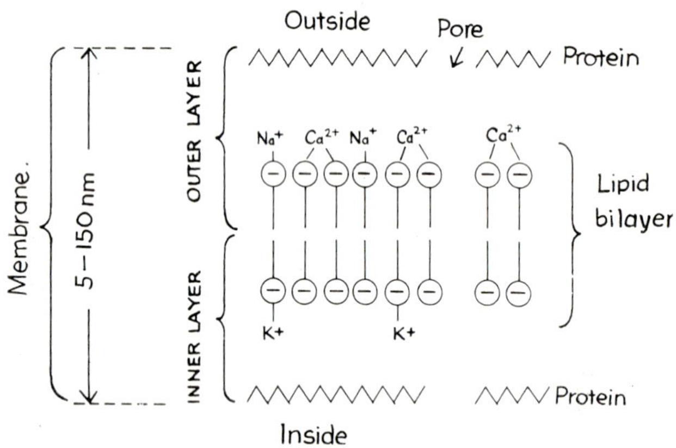

Fig. 9.1. Simplified representation of a typical biological membrane. The zig-zag lines are protein molecules which provide structural support ; the small circles are the hydrophilic (water-attracting) heads and the vertical lines are the hydrophobic (water-repelling but having an affinity for fatty molecules) tails of long lipo-proteins. Lipids are derived from long-chain organic acids or alcohols : fats, oils and waxes are lipids.

end-groups. The inside of the membrane is negatively charged with respect to its outside. A potential difference of about 50–100 mV (as the membrane is about 10 nm thick) exists across the membrane: this is the *membrane potential*. Thus, an electric field of the order of 10^7 V m^{-1} exists across the membrane.

The membrane may be excited in response to a stimulus, which may be electrical, mechanical, chemical, thermal or optical. The stimulus causes a series of electrochemical events: the membrane potential drops and may even, as in nerve membranes, invert so that the inside temporarily becomes positive with respect to the outside, and the permeability of the membrane to various ions and molecules alters.

The membrane acts as a complex electrode system involving a *double* double layer. The movement of ions, and their interaction with the local field, govern the behaviour of and events at the membrane surface.

A modern electrodic view of membrane potentials was put forward by Del Duca and Fuscoe in 1965, and further developed by Lowenheim in 1968. According to them, the membrane potential arises in a way analogous to the steady state potential of a fuel cell; see fig. 9.2.

9.3. *Energetics*

While a green plant is capable of utilizing sunlight directly via the processes of photosynthesis, animals and non-photosynthetic plants require another source of energy: food. A man consumes some 12 500 kJ per day as food, so that he ingests energy at an average rate of about 160 W (smoothing out his intake over the day). In heat engines, the thermodynamic conversion efficiency depends on the temperature differential between the hot part and the cold part of the thermal-to-mechanical energy conversion process. It is given by:

$$\epsilon = (T_{max} - T_{min})/T_{max}. \tag{9.1}$$

Now, an animal cannot support temperature differences greater than about 5 K. Thus, ϵ becomes about 1·5 per cent so that the human system would have at its disposal something like $160 \times 1{\cdot}5/100 = 3{\cdot}2$ W. But the human circulatory system alone requires about 10 W. We thus must look for a mechanism of energy conversion which is not limited by the thermodynamics of a heat engine; and we have seen in Chapter 5 that electrochemical energy conversion, as exemplified in the fuel cell, is such a way.

What is the nature of the electrochemical engine in biology and what are the reactions involved? Let us look at the electrode potentials of the known biochemical reactions involved in the process of extracting energy from food, which is digested by being oxidized in a series of linked redox reactions (Table 9.1). Biochemical reactions involve biochemical catalysts called *enzymes*. An enzyme is specific; it mediates just one step in a complex chain of reactions. Take as an

Electrode couple	Enzyme	Potential V_0 /V
acetate/acetaldehyde	xanthine oxidase	−0·58
lactic acid/xanthine	xanthine oxidase	−0·39
gluconolactone/glucose	glucose oxidase	−0·36
cystine/cysteine	none	−0·33
acetaldehyde/ethanol	alcohol dehydrogenase	−0·20
pyruvate/lactate	lactic dehydrogenase	−0·18
oxaloacetate/malate	malic dehydrogenase	−0·16
fumarate/succinate	succinic dehydrogenase	+0·02
dehydroascorbate/ascorbate	ascorbic oxidase	+0·06
ferricytochrome c/ferrocytochrome c		+0·27
oxidized cyt. oxidase/cytochrome oxidase		+0·29
$O_2+4H^++4e/2H_2O$		+0·82

(After M. G. Del Duca and T. M. Fuscoe, *Sci. and Tech.*, **1965**, p. 56.)

Table 9.1. Reversible electrode potentials (at pH 7) of biochemical reactions.

example the oxidation of glucose (G) to gluconolactone (GL) by oxygen in the presence of the enzyme glucose-oxidase, GO :

Firstly, protons are discharged on to the enzyme which is thereby reduced :

$$G+GO+2H^++2e \rightarrow (GO)H_2+GL. \tag{9.2}$$

Then oxygen is electrochemically reduced on the enzyme, forming H_2O_2, and the enzyme is thereby oxidized back to the starting point :

$$(GO)H_2+O_2 \rightarrow GO+H_2O_2. \tag{9.3}$$

We thus have an oxidation-reduction (or redox) system which is characterized by its reversible potential.

The energy available for effecting some reaction can be calculated as the difference between the lowest and highest redox potentials in the chain, ΔE. Thereafter, to obtain the available free energy, we apply $\Delta G=-zF\Delta E$. However, biologically, the overall free energy change is being carried out in several steps, each involving little energy change.

The series of reactions associated with the oxidation of glucose could be regarded as constituting the anodic reaction in a fuel cell, taking place in a region of a membrane. At other sites on the membrane, oxygen is being reduced in a cathodic reaction. (We might call these sites *anode* and *cathode* respectively, since the resulting system resembles a fuel cell.) Since the medium inside a living system is an aqueous electrolyte, ions travel freely in it between the sites.

There remains the question of the return path for the current, corresponding to the external-circuit part of a fuel cell circuit. Many

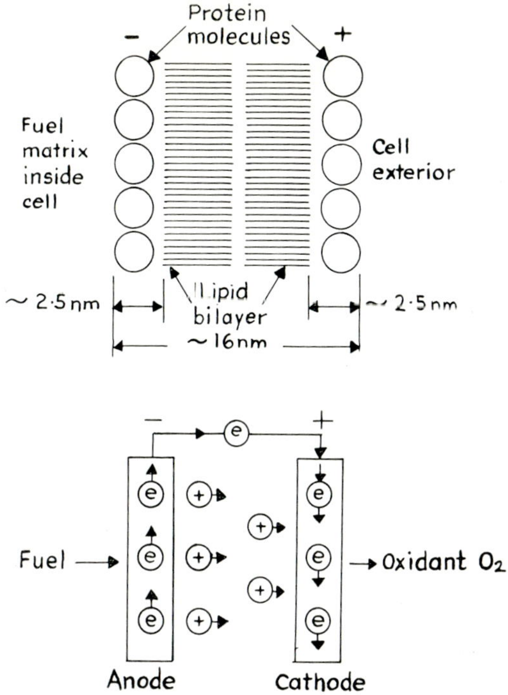

Fig. 9.2. Del Duca and Fuscoe's model illustrating how the cellular membrane resembles an electrochemical fuel cell. After M. G. Del Duca and J. M. Fuscoe, *Sci. and Tech.*, **1965**, 56.

organic materials exhibit a surprising degree of electronic conductivity. and it is by a flow of electrons through the solid biological materials that the cell circuit is completed.

We thus end up with the microscopic fuel cell across a membrane as illustrated schematically in fig. 9.2. Fuel is supplied to the inside and oxygen, or other oxidant, to the outside and the reaction proceeds as indicated.

The sites of the bio-electrochemical reactions are on the membrane, which has recently been demonstrated to have enough electrical conductivity to allow a metal to be electroplated on its surface.

It is the electrochemical nature of the electrocatalytically promoted reactions which results in the wizardry of the cell's biochemistry, which can carry out the most complex reactions at room temperature. As we have seen in Section 5.3, p. 79, and Section 4.6, p. 61, a change of potential from the reversible value increases the rate of the reaction exponentially without the use of a high temperature; 100 mV change in overpotential changes the rate of an electrochemical reaction by a factor of ten (see Chapter 4, p. 66). Lowenhaupt has termed this potential difference the *bio-electrochemical throttle of metabolism.*

Evidence that, at least in some biological systems, these electrochemical reactions are rate-controlling, has been adduced by Mandel who has shown that biological reactions on the membrane surfaces which he studied not only obeyed an equation of the form of the electrochemical rate equation (equation (5.12)), but that the Tafel slope of these reactions had values corresponding to a symmetry factor $\beta \simeq 0{\cdot}4$ to $0{\cdot}5$. Thus, the membrane is not merely a structural support for the enzyme systems in their proper sequence, but provides the essential interface between two regions held at different electrochemical potentials ; as Mandel says ' it is a universal component of biological electrochemical energy conversions '. According to Mandel, it is the discharge of H^+ ions across the membrane-solution interface which is the elementary rate determining step, and which controls the overall biological process.

The quantum aspects of proton transfer at electrodes (here, at membranes) have been applied by Lowdin to ageing and to the formation of some types of cancers. All these have in common a failure of the bio-electrochemical throttle of metabolism.

Thus, according to this view, membrane potentials arise from differences of the ' mixed ' potentials at which the cathodic and anodic reactions are being carried out at equal and opposite rates on each side of the membrane.

Na^+ and K^+ are transported across the membrane during the functioning of the cell. The passage of these ions is not—as used to be thought—the cause of the potential which can be measured across the membrane. They are transported in a simple electrochemical cell, as a function of the passage of current between the two sides of the membrane.

9.4. *A few examples of bio-electrical effects*

9.4.1. *Nerve signal transmission*

A special case of a membrane potential is the *action potential* in which an electric current is transmitted along a nerve (an *axon*). The potential spike produced is illustrated in fig. 9.3 (*a*), together with the changes in the membrane permeability to sodium and to potassium ions which accompany the action potential. The disturbance travels along the axon giving rise to local current (fig. 9.3 (*b*)). The transmembrane currents are illustrated (fig. 9.3 (*c*)) : among the events which occur is the electrodiffusion of K^+ which moves out while Na^+ moves in.

In the steady state, there is a Na^+ deficiency inside and a K^+ excess relative to the outside ; when the action potential occurs, it inverts this concentration gradient. To re-establish the initial state, an older theory of membrane potentials requires an imaginary ' sodium pump ' which pumps the Na^+ out of the axon and the K^+ back into it. In spite of many decades of research, no evidence for the actual existence of such a device has been found.

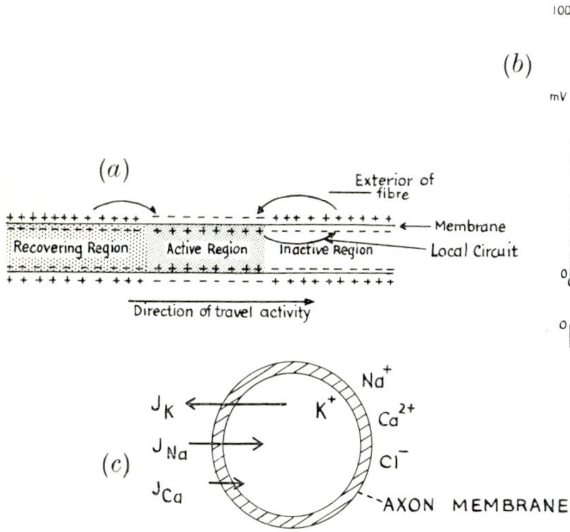

Fig. 9.3. The action potential of a nerve fibre. (*a*) A current stimulates the nerve, producing the action potential, V_m. The associated changes in the conductance for Na^+ and K^+, (g_{Na} and g_K respectively) are also shown. The right-hand scale refers to conductance changes and the ordinates on the left to V_m. (*b*) Diagram of the events during the propagation of the ' action potential ' spike, and creation of the current in the nerve axon. The reversal of the membrane potential produces the local current between active, and as yet inactive, regions and excitation of the new spike along the axon, with its propagation into a more distant, as yet inactive, region. After J. O'M. Bockris and D. M. Drazic, *Electrochemical Science*, Taylor and Francis, London, 1972. (*c*) Cross-section diagram of a squid axon showing the ionic fluxes associated with the action potential. The total inward membrane current I is due to the fluxes J_{Na}, J_{Ca} and J_K as indicated. After I. Tasaki and I. Singer, *Ann. N.Y. Acad. Sci.*, **137**, 792 (1966).

According to the electrodic view, there is no need for the elusive sodium pump†. A potential difference causes electrode reactions to occur, and hence a current passes ; i.e. $K^+ + Na^+$ ions are transported. The primary process envisaged is an electrodic reaction involving H^+,

† In green plants, action potentials may be propagated even in Na-free distilled water. The sodium pump concept thus cannot be applicable there.

the membrane acting again as the electrode, followed by an exchange of the protons for K^+ ions, yielding the movement of K^+ observed as in a normal electrochemical cell. Instead of the sodium pump, the excess H^+ ions become localized at the negative membrane sites (fig. 9.1) and repel Na^+ ions, so that the sodium current is reduced.

These hypotheses certainly require further elaboration ; however, they are beginning to offer an alternative to the sodium pump hypothesis.

9.4.2. *Electro-encephalography*

Continuous electrical oscillations can be recorded from the nervous system. The amplitude and frequency depend on the region of the brain which is being observed.

This effect is used clinically in electro-encephalography (EEG), in which 8 electrodes are attached to the scalp of a subject and the several resulting patterns recorded. The electrochemistry of neural processes is a mechanistically unexplored field.

9.4.3. *Bone growth*

Bone is about 45 per cent crystalline hydrated calcium phosphate (a mineral called apatite), plus some 30 per cent of fibres, composed of a protein called collagen. The rest is water. The apatite crystals are arranged in an orderly fashion relative to the collagen fibres and the collagen–apatite interfaces behave like the interface between two types of semiconductors (say, the p–n junction between p-type and n-type germanium) which can be made by adding electron-donating entities to one side and electron-accepting entities to the other.

Becker and Bachmann have shown that mechanical stress applied to bone produces electrical potentials and currents, which are associated with these p–n junctions. Conversely, the application of external electric fields affects the healing of bone lesions : if the field is applied across the lesion, collagen fibres are forced into alignment and the healing process is speeded up. If the field is applied perpendicularly to that direction, new tissue grows in a disorganized manner and healing is retarded.

Currents of electrochemical origin result whenever living bone tissue undergoes a lesion. Becker and Bachmann have suggested that these furnish a signal to the apatite/collagen junction matrix, which stimulates the repair mechanism. If these internal currents are augmented by correctly applied external fields, the process is speeded up. It is the potential which is important ; nature originates it, probably electrochemically, but man can help with a battery. Thus, in clinical trials by Brighton, broken bone structures which were reluctant to heal have been treated successfully by connecting an external battery across the fracture ; it is as if new bone material were being ' electrodeposited '.

One is reminded, too, of the electrophoretic deposition process of colloidal rubber, or of colloidal paint particles.

Teeth likewise exhibit a bone-like structure and Bevilacqua has recently shown that mechanical stress generates electric potentials in tooth structures. It may be that the mechanism of tooth decay involves aspects which are like those of electrochemical corrosion.

9.4.4. *Blood clotting*

The walls of blood vessels are biological membranes ; their internal surface (' intima ') is negatively charged, as shown by Sawyer and Srinivasan. These workers have demonstrated that the blood clotting, or thrombus formation, process is electrochemical in nature because :

(*a*) Metals inserted into the blood stream cause clotting if their potential in the electrochemical series (see p. 115) is positive with respect to the normal hydrogen electrode, and are free of thrombus formation if their potential is negative in the series.

(*b*) If a metal wire is inserted into the blood stream, blood coagulation can be turned on and off by varying the potential of the wire with respect to the blood : a negative wire potential inhibits and a positive one promotes coagulation.

(*c*) The mass of the clot formed appears to obey Faraday's law increasing with the passage of the current.

Blood is a complex colloid electrolyte and thus its component particles are electrically charged ; the interaction between the charged blood particles and the intima thus is an interfacial electrochemical reaction, the intima behaving as an electrode, as do other biological membranes.

These studies should lead to greater understanding of the mechanism of coronary thrombosis, apart from their bearing on the design of implants, pace-makers, etc.

CHAPTER 10

the electrochemical future

For the next few decades, the environmental aspects of life may be the chief areas which will attract scientists to electrochemistry and electrochemical engineering. Electrochemical technology offers solutions—in some cases the only ones foreseeable at present—for the maintenance of a world which can support the present numbers, without a great lowering of living standards or population.

Electrochemical interfacial reactions are already everywhere in practical life. They underlie the operation of batteries, fuel cells, the extraction of metals, desalination, electrosynthesis, corrosion, pharmacological action, etc.

Among the advances to be expected from electrochemical technology will be a cheap photogalvanic device for the conversion of solar energy to electricity ; the application of electrochemistry to the body, perhaps a fuel-cell driven heart ; non-polluting electrochemical syntheses ; non-polluting re-cycling of metals ; and the production of significant additional supplies of fresh water as a by-product of a hydrogen economy.

Future towns may be run on hydrogen and electricity. A town would be self-contained with all processes run electrically. Those involving matter will, therefore, be run electrochemically. Moving vehicles will run on fuel cells. Materials will be downgraded during use and regenerated electrochemically. Polluted liquids, including sewage, will be electrochemically purified. Cheap heat will be available from batteries charged during off-peak times. Electrometallurgical extraction of materials on a large scale from the moon is a possibility. In electrochemical biology, many speculative possibilities can be seen ; eventually, perhaps, communication from the brain by means of electrochemical mechanisms to artifical limbs, and from the brain directly to circuits not physically connected to the body.

INDEX

THE WYKEHAM SCIENCE SERIES

1 *Elementary Science of Metals* — J. W. MARTIN and R. A. HULL
2 *Neutron Physics* — G. E. BACON and G. R. NOAKES
3 *Essentials of Meteorology* — D. H. MCINTOSH, A. S. THOM and V. T. SAUNDERS
(*Paper and Cloth Editions available*)
4 *Nuclear Fusion* — H. R. HULME and A. MCB. COLLIEU
5 *Water Waves* — N. F. BARBER and G. GHEY
6 *Gravity and the Earth* — A. H. COOK and V. T. SAUNDERS
7 *Relativity and High Energy Physics* — W. G. V. ROSSER and R. K. MCCULLOCH
8 *The Method of Science* — R. HARRÉ and D. G. F. EASTWOOD
9 *Introduction to Polymer Science* — L. R. G. TRELOAR and W. F. ARCHENHOLD
10 *The Stars; their structure and evolution* — R. J. TAYLER and A. S. EVEREST
(*Paper and Cloth Editions available*)
11 *Superconductivity* — A. W. B. TAYLOR and G. R. NOAKES
12 *Neutrinos* — G. M. LEWIS and G. A. WHEATLEY
13 *Crystals and X-rays* — H. S. LIPSON and R. M. LEE
14 *Biological Effects of Radiation* — J. E. COGGLE and G. R. NOAKES
(*Paper and Cloth Editions available*)
15 *Units and Standards for Electromagnetism* — P. VIGOUREUX and R. A. R. TRICKER
16 *The Inert Gases: Model Systems for Science* — B. L .SMITH and J. P. WEBB
17 *Thin Films* — K. D. LEAVER, B. N. CHAPMAN and H. T. RICHARDS
18 *Elementary Experiments with Lasers* — G. WRIGHT and G. FOXCROFT
19 *Production, Pollution, Protection* — W. B. YAPP and M. I. SMITH
(*Paper and Cloth Editions available*)
20 *Solid State Electronic Devices* — D. V. MORGAN, M. J. HOWES and J. SUTCLIFFE
21 *Strong Materials* — J. W. MARTIN and R. A. HULL
22 *Elementary Quantum Mechanics* — SIR NEVILL MOTT and M. BERRY
(*Paper and Cloth Editions available*)
23 *The Origin of the Chemical Elements* — R. J. TAYLER and A. S. EVEREST
24 *The Physical Properties of Glass* — D. G. HOLLOWAY and D. A. TAWNEY
25 *Amphibians* — J. F. D. FRAZER and O. H. FRAZER
26 *The Senses of Animals* — E. T. BURTT and A. PRINGLE
27 *Temperature Regulation* — S. A. RICHARDS and P. S. FIELDEN
(*Paper and Cloth Editions available*)
28 *Chemical Engineering in Practice* — G. NONHEBEL and M. BERRY
(*Paper and Cloth Editions available*)
29 *An Introduction to Electrochemical Science* — J. O'M. BOCKRIS, N. BONCIOCAT, F. GUTMANN and M. BERRY

THE WYKEHAM TECHNOLOGY SERIES

1 *Frequency Conversion* — J. THOMSON, W. E. TURK and M. J. BEESLEY
2 *Electrical Measuring Instruments* — E. HANDSCOMBE
3 *Industrial Radiology Techniques* — R. HALMSHAW
4 *Understanding and Measuring Vibrations* — R. H. WALLACE

All orders and requests for inspection copies should be sent to the appropriate agents. A list of agents and their territories is given on the verso of the title page of this book.